DISTRIBUTED COMPUTING

MODELS AND THEORIES

DISTRIBUTED COMPUTING

MODELS AND THEORIES

Dr. Raj Gaurang Tiwari
Dr. Ambuj Kumar Agarwal
Dr. Abeer Aljohani
Dr. Rajat Bhardwaj

NAMYA PRESS.com

Published by

Namya Press

USA: 10685-B Hazelhurst Dr. # 31013 Houston, Texas USA - 77043
Email: namyapress@gmail.com
Website: https://namyapress.com

Edition	:	First Published in 2023 (Namya Press)
Title	:	Distributed Computing: Models and Theories
Authors	:	Dr. Raj Gaurang Tiwari, Dr. Ambuj Kumar Agarwal Dr. Abeer Aljohani & Dr. Rajat Bhardwaj
ISBN	:	978-1-956861-30-3
Copyright	:	© Dr. Raj Gaurang Tiwari, Dr. Ambuj Kumar Agarwal Dr. Abeer Aljohani & Dr. Rajat Bhardwaj. 2023 All Rights Reserved

Preface

The development of distributed systems in the recent years has been nothing short of astonishing. This expansion may be attributed to a number of factors, including the decreasing price of hardware, improvements in communication technologies, the meteoric rise of the Internet, and our growing reliance on networks for everything from social interaction to monetary transactions. New areas of application, such sensor networks and wearable computers, have been made possible by developments in embedded systems, nanotechnology, and wireless communication. The landscape of distributed computing has shifted as a result of the fast development of cloud computing and the increasing significance of big data.

Despite these systems' pervasiveness, they are often designed and developed by a mysterious cabal of experts. Distributed systems, however, are developing, standardising, and abstracting at the same rate as the rest of technology.

Communication, replication, fault tolerance, and security are just few of the topics this book delves into at length. It also covers the fundamentals of distributed systems and more advanced ideas and technology related to them. It was written for professionals and advanced students taking a course on distributed systems. It provides a thorough overview of the process by which distributed systems in the real world are developed and deployed.

DESCRIPTION, APPROACH AND FEATURES

The core concepts and underlying models of distributed computing will be the subject of this book. The course will cover the fundamentals of distributed computing, including its theory, techniques, and underlying systems. Algorithms are presented in a straightforward fashion, with the core concepts and intuition being explained using diagrams and easy explanations rather than being buried behind frightening notations and long and difficult to understand formal proofs of the algorithms. The book's subject selection is wide and inclusive, and it discusses every significant issue at length. Each chapter's algorithms were chosen with attention so that they would best illustrate the most relevant and cutting-edge aspects of algorithm creation. This book presents the theory behind and algorithms for a variety of practical system-like problems (such as mutual exclusion, deadlock detection, termination detection, failure recovery, authentication, global state and time, etc.), despite a primary focus on the fundamentals of distributed computing. The book is prepared with the effect of new trends, such as peer-to-peer computing and network security, on the basics of distributed computing in mind.

There are illustrations, cases, examples, a practise problem, a summary, and resources for further reading in each chapter.

READERSHIP

This book is aimed as a textbook for the following:

- Graduate students and Senior level undergraduate students in computer sience and computer engineering.
- Graduate students in electrical engineering and mathematics. As wireless networks, peer-to-peer networks and mobile computing continue to grow in importance, an increasing number of students from electrical engineering departments will also find this book necessary.
- Practitioners, system designers/programmers, and consultant in industry and research laboratories will find the book a very useful reference because it contains state-of-the-art algorithms and principles to address various design issue in distributed systems, as well as the latest references.

PERQUISITES

Hard and soft perquisites for the use of this book include the following:

- An undergraduate course in algorithms is required.

- Undergraduate courses in operating systems and computer networks would be useful.

READER FEEDBACK

Feedback from our readers is always welcome. Let us know what you think about this book—what you liked or disliked. Reader feedback is important for us as it helps us develop titles that you will really get the most out of.

To send us general feedback, simply e-mail <rajgaurang@gmail.com>, and mention the book's title in the subject of your message.

QUESTIONS

If you have a problem with any aspect of this book, you can contact us at <rajgaurang@gmail.com>, and we will do our best to address the problem.

—Editors

Contents

1

Operating System Basics

1.1 OVERVIEW

A modern computer system consist of one or more processor main memory, clocks, terminals, disks n/w interfaces and other input output devices. Writing programs to keep track of all these components and use them efficiently and effectively is extremely difficult job. If a programmer has to be concerned about how data is read or written on the disk, it is unlikely that any program could be written at all. So it is clear that some mechanism had to be designed to shield programmers from complexity of the hardware.

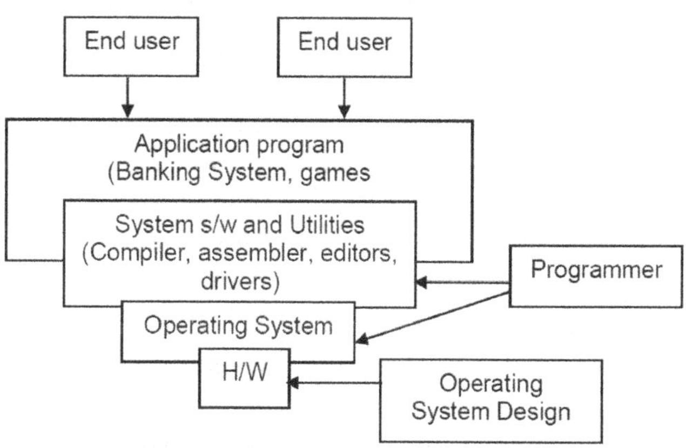

Figure 1.1: Operating system Layer

The mechanism that has gradually evolved is to put a layer of software on the top of n/w to manage all part of system and present the user with an interface that is user with an interface that is easier to understand and program. This layer of software is the operating system (Figure 1.1).

1.1.1 Layers & View of Computer System

An operating system conceals the truth about n/w from user. It performs all low level functions that are required to lead & execute a user's program. Programmers now need not to worry about the details of writing data into the disk, reading information from disk and store them in a memory location.

- Operating System Services
- Interprets and executes user command
- Manages the system h/w
- Allocates system resources
- Schedules process
- Facilitates recovery
- Maintains file system
- Performs administrative functions
- Performs house keeping task.

1.1.2 Hierarchical Operating System Structure

In a strictly hierarchical implementation a given level is allowed to call upon services of lower level, but not on those of higher levels.

What functions should be place in the lowest level (level 1). The one that is used by all resources manager is keeping track of the allocation of resources. This required synchronization of resource allocation. The primitive operations which operate upon a s/w mechanism called semaphore are:

- P operator (noting that resource is seized or allocated)
- V operator (noting that resource has been released)

A semaphore is associated with each resource in the system.

When a resource is requested, the P operator is used to test the corresponding semaphore, if it is off P turns it on and returns. If semaphore is already on the P. P makes a note that requesting process has been placed in wait state waiting for resource associated with that semaphore. Later, V operator issued by another process releases the resource, turns the semaphore on and awakens any process waiting for that resource.

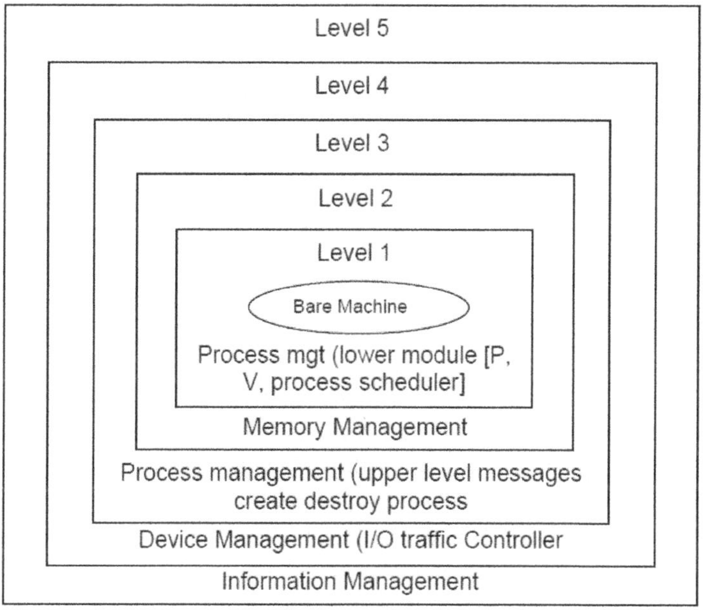

Figure 1.2: Layered architecture of Operating System

Level 1	*Processor Management lower level*
	1. P Synchronization primitive
	2. V Synchronization primitive
	3. Process scheduling (Mechanism of multiprogramming
Level 2	*Memory Management*
	1. Allocate memory
	2. Release memory
Level 3	*Processor management Upper level*
	1. Create/destroy process
	2. Send/receive message between process
	3. Stop process

Level 4	*Device management*
	1. Keep track of all I/O device
	2. Schedule I/O
	3. Initiate I/O process
Level 5	*Information management*
	1. Create/destroy file
	2. Open/close file
	3. Read/write file

1.2 EVOLUTION OF OPERATING SYSTEMS

In attempting to understand the key requirements of an operating system and the significance of major features of an OS it is useful to understand how operating system have evolved over the years.

1.2.1 Serial Processing

With the earliest computers, the programmers interacted directly with the computer h/w (there was no operating system)

Programs for the bare machine can be developed by manually translating sequences of instruction into binary code. Instructions & data are then interred into computer by means of console switches, or perhaps through a hexadecimal keyboard. Programs are started by loading the program counter with the address of first instruction. Results of execution are obtained by examining the content of relevant registers and memory locations. I/O devices if any must be controlled by executing the program directly.

In a typical sequence, the editor program is loaded to prepare the source code of user program, next step is to load and execute language translator and to provide it with source code of user program. If the errors are detected, the whole process must be repeated from beginning. After that object code produced from syntactically correct source code is loaded and executed. If run time errors are detected, an assistant program called debugger is loaded.

1.2.1 Batch Processing

The next logical step in the evolution of operating system was to automate the sequencing of operations involved in execution and in mechanical aspect of program development.

The common input devices at that time were card reader and tape drives and common output devices were line printer, tape drives and card punches.

In the batch processing, the user do not have direct access to machine. Rather he submits the job to a computer operator, who batches the jobs together with similar needs sequentially and places the entire batches on an input device for the use by a piece of software known as monitor. Memory Layout in Batch Systems is depicted in figure 1.3.

It is the monitor that controls the sequence of events, means as soon as a job completes its execution; it automatically starts loading the next job.

Monitor always resides in main memory and available for execution. This portion is referred as resident monitor.

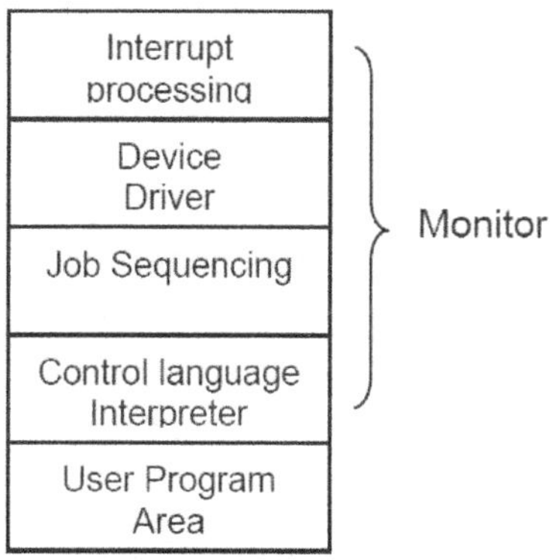

Figure 1.3: Memory Layout in Batch Systems

The monitor reads a job one at time from input device, as it is read in, it is placed in user program area and control is passed to this job. When the job is completed it returns control to monitor which immediately reads in next job. The result of each job is printed out for delivery to user.

How a Job Differs from Program

Some instructions are included with user program and data to form a job. These instructions are written in Job Control Language (JCL) which is a special type of language used to provide instructions to monitor. Typical JCL commands include making of job beginning and end, command for loading and execution of program, and command to announce resource needs such as expected execution time, and memory allocation. When a JOB_END command is encountered, the monitor may look for another job which may be identifies by a JOB_START command.

Advantage

- Reduces the CPU idle time due to slow manual operation.
- Offers greater through put than simple serial processing.

Disadvantage

- As far as program development to concerned batch is not a good improvement over simple serial processing.
- Turned around time (measured from submission of job to output received) is quite long.
- Programmers are forced to debug their programs off line using post mortem memory dumps as apposed to being able to examine the state of machine immediately upon detection of failure.
- Speeds of mechanical Input Output devices are much slower than CPU (an electronic device). This also a major performance bottleneck.

1.2.3 Multiprogrammed Operating System

After batch operating systems, to increase the system throughput and resource utilization a class of operating system that multiplex resources of computer system among multiple active programs were evaluated. They are:

- **Multitasking:** An instance of program in execution is called a process. A multitasking operating system is distinguished by its ability to support the concurrent execution of two or more active processes. Multitasking is usually implemented by maintaining code and data of several processes in memory simultaneously and by multiplexing processor and Input Output devices among them. As soon as a program being executed encounters any I/O instruction and, it needs wait for I/O to be completed, the processor switch to another job.

- **Multiprogramming:** The term multiprogramming denotes an operating system that, in addition to supporting multitasking, provides sophisticated forms of memory protection and enforces concurrency control when process access shared I/O devices and files. Multiprogramming OS usually support multiple users.
- **Multiuser:** Multiuser operating systems provide facilities for maintenance of individual user environments require user authentication for security and protection and provide per-user resource usage accounting.

Note:

- In general, multiprogramming implies multitasking but multitasking does not imply multiprogramming.
- Multitasking is one of the mechanisms that a multiprogramming operating system employs in managing the totality of computer system resources.
- Multiprogramming with two programs.

1.2.4 Time Sharing Operating System

- Time sharing operating system is logical extension to multiprogramming. CPU executes multiple jobs by switching among them, but the switches occur so frequently that the users can interact with each program while it is running.
- Time sharing OS allows many users to share the computer system simultaneously. Every user program is allocated a small period of CPU time (typically 10 to 20 milliseconds) known as time slice.
- As system switches rapidly from one user to another user each user is given impression that the entire computer system is dedicated to him, even though it is being shared among users.

1.2.5 Real Time System

It is special purpose operating system. It is used when rigid time requirements have been placed on the operation of processor or data flow. It is used as control device in a dedicated system.

A real time system has well defined, fixed time constraints. Processing must be done within defined constraints otherwise system will fail. That means a real time system, functions correctly only if it returns the correct result within its time constraints.

Thus real time operating system are used in environments where large number of events, mostly external to computer system, must be accepted and processed in a short time or within certain deadlines. Such applications include industrial control, telephone switching equipment, weather forecasting, flight control etc.

A primary objective of real time system is to provide quick event response time and thus meet the scheduling deadlines. User convenience and resource utilization are of secondary concern to real time system designers.

1.2.6 Multiprocessor System

There are basically two ways to increase the speed of computer system:

- By using high speed components architecturally. Such components tend to be expensive and are often less reliable than standard.
- By introducing multiprocessor system which provide an appealing architecture alternative for improving performs of computer systems by coupling a number of low cost standard processors.

Thus multiprocessor system has more than one processor in close communication, sharing the computer bus, clock and sometimes memory and devices.

Multiprocessing can be applied to provide:

1. **Increased system throughput** by executing a number of different user process on different processors in parallel.
2. **Application speedup** by executing some portion of the application in parallel.

   Note:

   - Throughput can be improved by executing a number of unrelated user process on different processes in parallel. Thus throughput is improved by completing large number of tasks in a unit time.

- Application speedup may be obtained by exploiting parallelism within an application and by creating multiple processes or threads that may be schedule for execution on different processes.

3. **Economy of Scale:** Multiprocessors systems can save more than single processors systems, because they can share peripherals, mass storage and power supplies.
4. **Fault Tolerance:** If functions can be distributed properly among several processors, then failure of one processor, will not halt the system, but it will be just slow. If we have to processors and one fails then remaining nine processors will share the work of failed processor. Thus the entire system will run 10% slower rather than failing altogether.

   Note:

   - The speed up ratio with N processors is not N, rather it is less than N. It is just because when multiple processors cooperate on a task, a certain amount of overhead is incurred in keeping all the parts working correctly. This overhead, plus contention for shared resources lowers the expected gain from additional processors.

Design Issues for Multiprocessor Systems

The most common multiprocessor systems use following policies (CSMP).

- **Symmetric Multiprocessing (SMP):** In this, each processor runs an identical copy of operating system, and these copies communicate with another as needed.
- **Asymmetric Multiprocessing (AMP):** In this each processor is assigned a specific task. A master processor controls the system. The other processors either look of master for instruction or have predefined task.

1.2.7 Distributed Systems

Distributed computer system is a collection of autonomous computer systems capable of communication and cooperation via their hardware and software interconnections. Distributed computer systems are generally characterized by:

- Absence of shared memory
- Unpredictable inter node communication delays.

In other words, distributed system is a collection of loosely coupled processor interconnected by a communication network. These processors are referred to by a number of names such as sites, nodes, computers, machines or hosts depending on the context in which they are mentioned.

A distributed OS governs the operation of distributed system and provides a virtual machine system abstraction to its user.

Advantage of Distributed System

- **Transparency:** Component and resource distribution is hidden from the users unless they explicitly demand it.
- Resource sharing and load balancing.
- Information sharing
- Reliability, availability and fault tolerance
- Communication

In the distributed OS, the users access remote resources in the same manner as they do local resources. Data and process migration from one site to another is under the control of distributed operating system.

1.2.8 Network Operating System

A network operating system provides an environment in which users who are aware of multiplicity of machines, can access remote resources by either logging into appropriate remote machine or transferring data from remote machine to their own machine.

1.3 OPERATING SYSTEM AS RESOURCE MANAGER

An operating system is a collection of programs designed to manage system resources which may be:

- Memory
- Processors
- I/O devices
- Information (programs and data)

All the resources are valuable and it is the function of OS to see that they are used efficiently and to resolve conflicts arising from competition among various users.

Operating system as a resource as a resource manager must do followings:

- Keep track of all resources
- Enforce policy that determine who gets what, when and how much.
- Allocates the resources
- Reclaim the resources

Thus the main services provided by OS can be grouped as:

- Memory management functions
- Processor management functions
- Device management functions
- Information management functions

1. Memory Management Functions

- Keep track of memory what parts are in the use and by whom? What parts are not in use (free)?
- If multiprogramming, decide which process gets it and how much
- Reclaim memory when process no longer needs it or has been terminated

2. Processor Management Functions

For this OS has following three programs:

- **Traffic Controller Processors:** Traffic controller keep track of processors and status of process.
- **Process Scheduler:** Process scheduler decides, chooses from all the jobs submitted to system and decides which one will have chance to use CPU.
- **Dispatcher:** Dispatcher allocates the processor by setting up necessary hardware registers.
- **Traffic Controller:** Traffic controller also reclaims the processor when process relinquishes processor or terminated or exceeds allowed time of usage.

3. Device Management Functions

- Keep track of devices, channels and control units (I/O traffic controller)
- Decide what is an efficient way to allocate a resource (device). Decides whether it is shared or dedicated device. In the case of shared devices, who gets it and how much. This function is also known as I/O scheduling.
- Allocate the resource (device) & initiate I/O operation.
- Reclaim resource in most cases I/O terminates automatically.

4. Information Management Function

- Keep track of resource (information), its location, use, status etc. These collective facilities are often called file system.
- Decide who gets the use of resources, enforce protection requirement and provide accessing routines.
- Allocate the resource (Information) i.e., open a file.
- De-allocate a resource i.e., close a file.

1.4 SYNCHRONIZATION MECHANISM

1.4.1 Concept of Process

Process is a program in execution. A process is more than the program code, which is sometimes knows as **text section**. It also includes current activity, as represented by value of **program counter** and the contents of processor's register. In addition, a process generally includes process **stack**, which contains temporary data (such as method parameters, return address and local variables) and a **data section**, which contains global variables.

Each process is represented in operating system by Process Control Block (PCB). It contains many pieces of information associated with it as shown in **figure 1.4**:

Pointer	Process State
Process Number	
Program Counter	
CPU Registers	
Memory Limits	
CPU Scheduling Info	
List of Open Files	
≡	

Figure 1.4: Process Control Block

1.4.2 Process State and Transition

As a process executes, it changes state. The state of a process is defined in part of current activity of that process. Each process may be in one of the following states (Figure 1.5):

- **Submit:** When we execute any program it is submitted to operating system as to what resources it will need. At this instance process is said to be in submit state.
- **Hold:** The SPOOLING routine reads the job and places it on to a disk. Now process waits for its turn in queue for some time. At this state process is said in hold state.
- **Ready State:** As the process advances in the queue to attain the CPU attention, scheduler scans the SPOOLED process, and call memory management to determine where sufficient main memory is available and also call device management whether the devices requested are available.

  Once the job scheduler decides that process is to be assigned resources, traffic controller is called to create associated process information, and memory management is called to allocate the necessary main storage. At this stage process is in ready state.
- **Run:** When processor becomes free, scheduler program scans the list of ready processes, chooses a process

and assigns processor to it. Now the process is called in run state. In the middle of execution it might happen that time slice allotted to this process gets over, then scheduler dispatch program puts this process again in ready state and assigns CPU to another process.

- **Wait:** Some process might require to do disk I/O. Since I/O is slow operation and CPU cannot lie idle till the time I/O is over. Therefore, such processes are put in wait state until their I/O is over. And as they complete their I/O is operation they are again put in ready.

  When the running process request to access an information (i.e., read a file), information management would call device management to initiate the reading of file. Device management would initiate I/O operation and then call process management to indicate that the process requesting the file is awaiting the completion of I/O operation.

  When I/O is completed, the I/O hardware sends a signal to traffic controller in process management, which places the process back into ready state.

- **Complete:** When the process completes its computation, it is placed in complete state and all resources are freed.

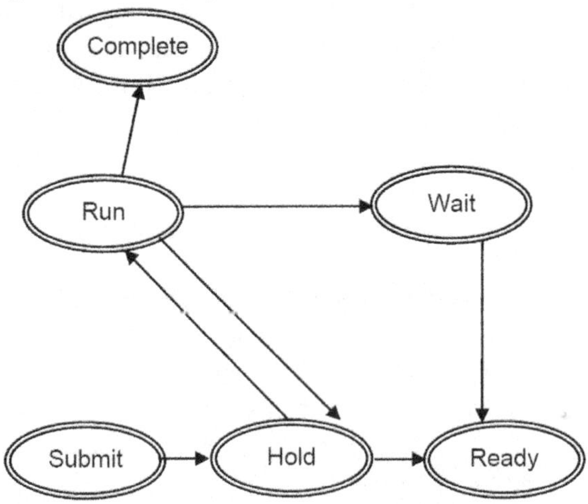

Figure 1.5: Process State Transition Diagram

Function of Operating system involved in handling diverse states of process is demonstrated in figure 1.6.

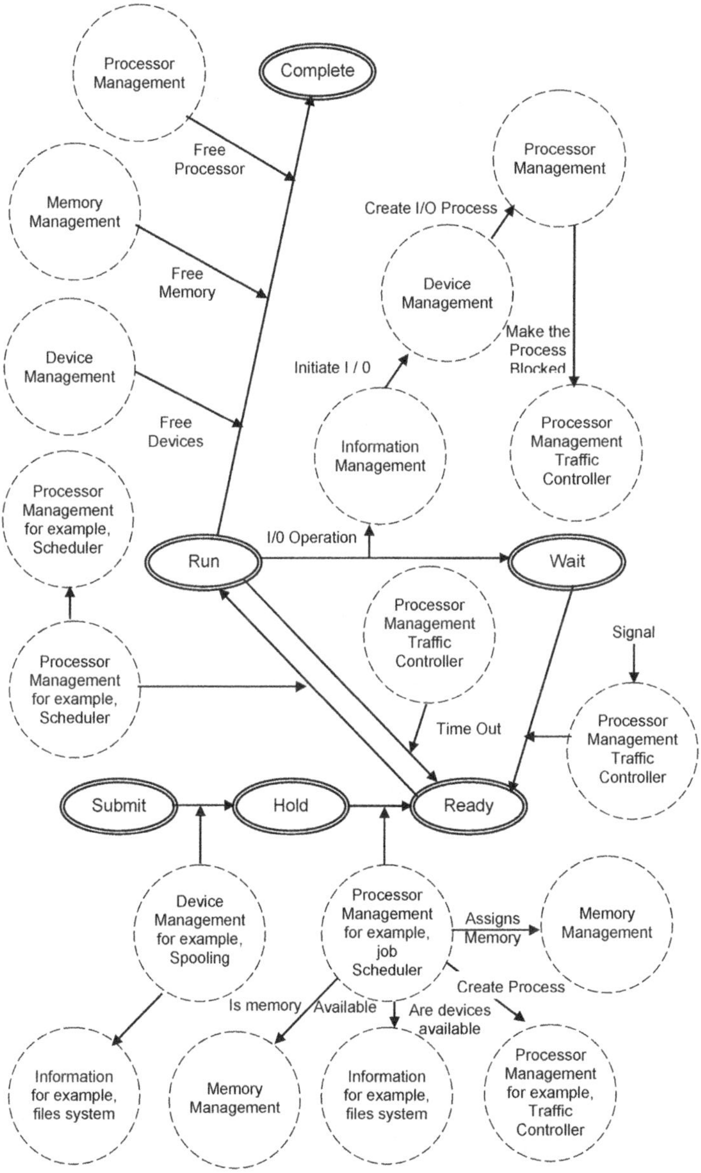

Figure 1.6: Operating System Handling of Process State

1.4.3 Context Switching

Switching the CPU to another process requires saving the state of old process and loading the saved state for new process. This task is known as context switching (Figure 1.7).

This context of process is represented in the PCB of a process (It includes the values of CPU registers, process state and memory management information. When context switch occurs, the Kernel saves the context of old process in its PCB and loads the saved context of new process scheduled to run. Context switching time is totally pure overhead because system does no useful work while switching.

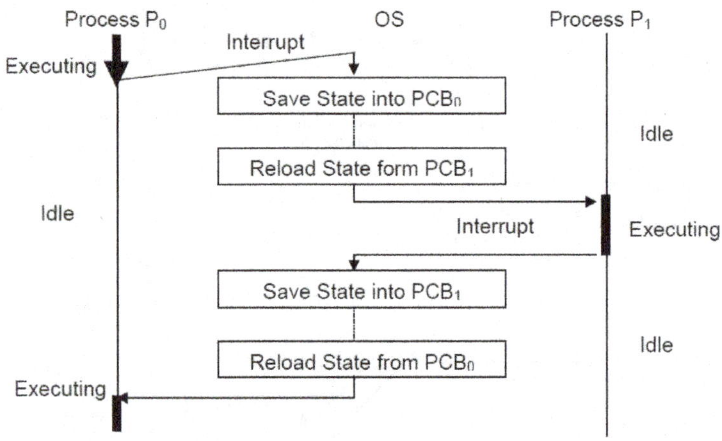

Figure 1.7: Context Switching

1.4.4 Principal of Concurrency

A general example to illustrate the concurrency may be typical rapid service burger stand (Figure 1.8). Customers submit your order to the order clerk. He places the order on order table. A packer takes the order from order table one at time and puts the requested item into a bag and delivers the completed order to the cashier.

The cashier processes these bags once at a time, receives money from customer and gives the order. Further more there is a cook working in back room, continually supplying the burger on a table.

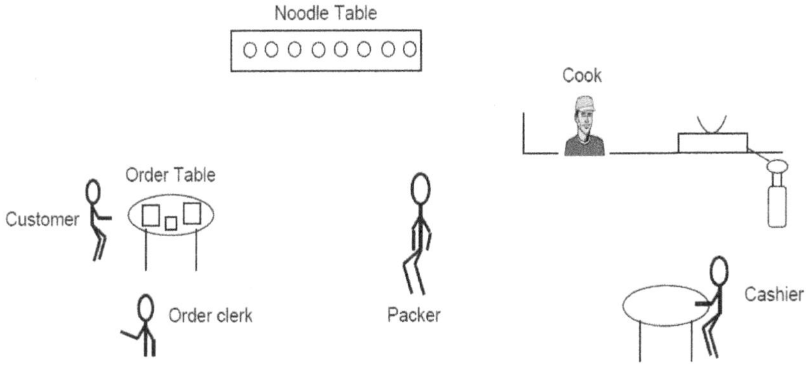

Figure 1.8: Example of Concurrency

In this example each process (person) operates independently. The only communication is via shared areas like order table and burger table. In a single processor multiprogramming system, processes are interleaved in time yield appearance of simultaneous execution.

1.4.5 Process Synchronization

A situation, where several processes access and manipulate the same data concurrently, and outcome of execution depends on particular order in which the access takes place, is called **race condition**.

Example of race condition in print spooler

When a process wants to print a file, it enters the file name in a special file named Spooler directory (Figure 1.9).

A process named Printer daemon periodically checks to see if there are any files to be printed, and it there are, it prints them and remove their names from spooler directory.

The prints daemon also maintains two shared variables.

- **Out:** Out points next file to be printed.
- **In:** In points next free slot in directory.

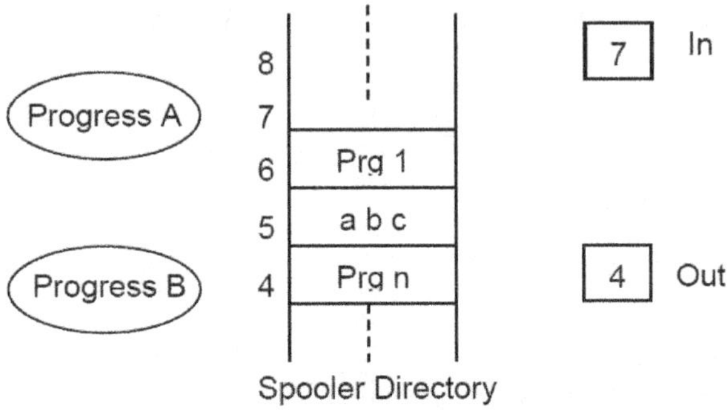

Figure 1.9: Spooler Directory Scenario

As soon as any process wants to print its file it executes following:

1. Read [in]
2. Store its value in local variable next_free_slot
3. Store the name of file in this slot
4. Increment the next_free_slot
5. Store it in [in].

Consider a situation, two processes A and B decide they want to queue a file in spooler directory for printing. The process A reads [in] and stores 7 in local variable next_free_slot. Just then a clock interrupt occurs and CPU decides process A has run long enough, so it switches to process B. Process B also reads [in] also gets 7 so it stores the file in slot 7 and update it with 8.

Eventually process A runs again, starting from the place it left off. It looks at next_free_slot finds 7 there and writes its file name in slot 7 erasing the name of process B an computes next_free_slot +1 which is 8 and sets it to [in].

Thus B never gets output

Thus when two or more processes are reading and writing some shared data final result depends on who runs precisely when, such a situation is race condition.

1.5 CONCURRENT PROCESS

Two processes are concurrent if their execution can overlap in time, i.e., the execution of second process starts before the first process completes.

- In multiprocessor system, different CPU's can execute different processes simultaneously, such concurrency is known as physical concurrency. This concept of concurrency is concrete and easy to visualize.
- In uniprocessor system, physical concurrency can be due to concurrent execution of the CPU and I/O. But the processes are interleaved in time to yield the appearance of simultaneous execution. Such a concept is known as logical concurrency.

1.5.1 Control Structure for Indicating Parallelism (PARBEGIN / PAREND)

If certain operations can logically be performed in parallel, there will be a significant improvement in performance. But it is difficult and time consuming to determine what activities can and cannot be performed in parallel. Many programming language constructs for indicating parallelism have appeared in the literature. These generally involve pairs of statements as follows:

1. One statement indicating that sequential execution is to split into several parallel execution sequences.
2. One statement indicating that certain parallel execution sequences are to merge and sequential execution is to resume.

These statements occur in pairs

- Parbegin / parend (for begin and end parallel execution)
- Cobegin / coend (for begin and end concurrent execution)

Parbegin / Parend Construct:

```
Parbegin
        Statement 1;
        Statement 2;
        — — —
        — —
        Statement n;
Parend
```

This Means:

Suspend the execution of main program; initiate concurrent execution of statement 1, statement 2, - - - statement n; when all of them have terminated, resume the main program.

Note: The term statement in parbegin/ parend construct may be simple statement, procedure call, block of sequential statement (enclosed by begin/end) or combination of these.

In other words, "When a program executing a single sequence of instructions, encounters parbegin construct, the single thread of control splits into n separate thread of control — one for each statement in parbegin/ parend construct. Each of the separate thread of control eventually terminates and reaches the parend, when all the parallel threads of control finally terminate, a single thread of control is resumed and system proceeds with the next statement after the parend.

Consider the following example:

```
x = (-b+(b**2 - 4*a*c)**.5) / (2*a)
```

This assignment might be evaluated on sequential processor as follows:

```
-b
b**2
4*a
(4*a)*c
(b**2) – (4*a*c)
(b **2 – 4*a*c) ** 0.5
(-b) + ( (b**2 - 4*a*c) ** 0.5)
2*a
(-b+(b**2 - 4*a*c) ** 0.5) / 2*a
```

Here nine operations are executed in sequence.

On the system that supports parallel processing expressions might be evaluated as follows:

```
parbegin
        temp 1 = -b
        temp 2 = b**2
        temp 3 = 4*a
        temp 4 = 2*a
parend
        temp 5 = temp 3 * c
        temp 5 = temp 2 – temp 5
        temp 5 = temp 5 ** 0.5
```

```
        temp 5 = temp 1 + temp 5
        x = temp 5 / temp 4
```

Another example:

```
For i = 1 to 100
        read a [ i ];
parbegin
        x = max (a);
        y = min (a);
parend
 result = x / y;
```

Question: The following expression describes the serial / parallel precedence relationships among six processes P_1 through P_6:

$P\ (S\ (P\ (P_3, S\ (P_1, P\ (P_6, P_5))), P_2, P_4)$

Where P indicates parallel and S indicates serial. Transform the expression into program using cobegin/ coend statement.

Solution:

```
Cobegin
        Begin
            cobegin
                P3;
                  begin
                      P1;
                        cobegin
                                P6;
                                P5;
                        coend
                     end
               coend
            P2;
         end
    P4;
coend
```

1.6 PROCESS SYNCHRONIZATION

To guard against the race condition we need to ensure that only one process at a time can be manipulator the shared thing. To make such a guarantee, synchronization of processes is required.

1.6.1 Critical Section Problem

The part of program where the shred memory, shared file or anything which is shared, is accessed is called **critical section.**

Each process has critical section. The important feature of system is that when one process is executing in its critical section, no other process is to be allowed to execute in its critical section. Thus the execution of critical sections by processes is mutually exclusive in time.

The critical section problem is to design a protocol that the process can use to co-operate.

1.6.2 Solution to Critical Section Problem

Each process must be request permission to enter its critical section. The section of code implementing this is the **entry section**. The critical section must be followed by an **exit section**. The remaining code is the **remainder section**.

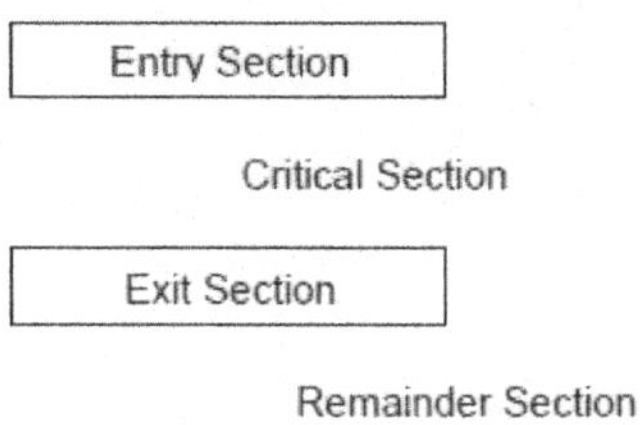

Figure 1.10: Structure of a Typical Process

Thus solution to critical section problem involves following three requirements:

- **Mutual Exclusion:** No two processes may be simultaneously inside critical section.
- **Progress:** No process running outside its critical section may block other process. Means for entry in the critical section only those processes will be considered who are

not running in remainder section. Processes who are running in remainder section and have wish to enter in critical section will not be considered.

- **Bounded Waiting:** No process should have to wait forever to enter its critical section.

1.6.3 Mechanism for Synchronization

For synchronizing the concurrent processes, some approaches which are used as follows:

Semaphore

The problem with the producer consumer is that when a wakeup is sent to a process who is not sleeping currently, it is lost. So to save this wakeup call, a wakeup waiting bit comes it to picture. It involves the concept. When a wakeup is sent to a process that is already awake, this bit is set. Later when the process tries to go to sleep, if the wakeup bit is on, it will be turned off, but process will stay awake.

Semaphore implements above concept. It is available which can be accessed only through following two operations:

- **Wait:** This operation on semaphore checks to see if the value is greater than zero, if so it decrements the value and just continues. If the value is zero, process is put to sleep and process is placed into waiting queue associated with semaphore.
- **Signal:** It increments the value of semaphore and if it fines it greater than 0, removes one process from list of waiting processes and awakens that process.

The implementation of semaphore is shown below;

```
Typedef struct
{
   int value;
   struct process * L;
   } semaphore;
```

Wait Operation

```
Void wait (semaphore S)
{
S. value ++;
```

```
if (S. value > 1)
{add this process to S.L;
sleep ()
   }
}
```

Signal Operation

```
Void signal (semaphore S)
{
   S. value - - ;
if (S. value >= 1)
{
   remove a process P from S. L;
   wake up (P);
   }
}
```

Mutual Exclusion Implementation with Semaphore

Each process, share a semaphore, mutex (stands for Mutual Exclusion)

The organization of process is shown below:

```
While (1)
{
   wait (mutex);
   critical section
   signal (mutex);
   non critical section
}
```

Path Expressions: Campbell and Habermann introduced the notation of path expressions to enable the specification of order in which procedures are to be executed.

Conceptually a 'path expression' is a quite different approach to process synchronization. A path expression restricts the set of admissible execution histories of operations on the shared resource so that no incorrect state is ever reached.

A path expression has the following form:

```
Path S end;
```

Where S denotes the possible execution histories. The operators used in path expression are:

1. Sequencing (;) : It defines the sequencing order among operations.

In the context of readers writers problem:

```
Path open; read; close; end;
```

Means an open must be performed first followed by a read and a close in that order. There is no concurrency in execution of these operations.

Note: When a path expression specifies a sequence of calls, it does not employ that the same caller must make each of calls.

2. Selection (+): It indicates only one of the operations connected by + operator can be executed at a time.

```
Path read + write end;
```

means only read or only write can be executed at same time.

3. Concurrency ({ }): It signifies that any number of instances of the operation delimited by { } can be in execution at a time.

For example:

```
path { read } end;
```

means any number of read operation can be executed simultaneously.

```
path write; { read } end;
```

either single write or several read operation be executed at same time. Between two write operations, at least one read operation must be executed. After all readers have finished, the writer must execute first.

```
path { read } + write end;
```

This is weak reader's priority solution. Due to {read}, if a reader is reading, subsequent readers gain immediate access to shared data. A waiting writer must wait until all readers have 'finished'. After a writer finishes, any one of the waiting writer or waiting readers can get access.

1.7 DEADLOCKS

1.7.1 Introduction

A system consists of finite number of resources to be distributed among a number of competing processes. The resources may be the processor I/O channel, main and secondary memory, devices and data structures such as files, databases and semaphores.

Under the normal mode of operation a process must request a resource after using it. A process may request as many resources as it requires to carry out its designated task. Thus a process may utilize a resource in only following way.

- **Request:** If the request can not be granted immediately, then the requesting process must wait until it becomes free.
- **Use:** The process can operate on resource.
- **Release:** Process releases the resource after performing the required task.
- **Note:** The number of resource requested may not exceed the total number resources available in the system.

A process in multiprogramming environment is said in a state of deadlock if it is waiting for a particular event that will never occur. The event with which we are mainly concerned here are resource acquisition and release. Thus:

"a process request resources, and if the resources are not available at that time, the process enters a wait state. It may happen that waiting process will never again change state, because the resources it has requested are held by other waiting process. This situation is called deadlock."

A system is said to be in deadlock if one or more processes are deadlocked.

Consider the following scenario shown in figure 1.11a:

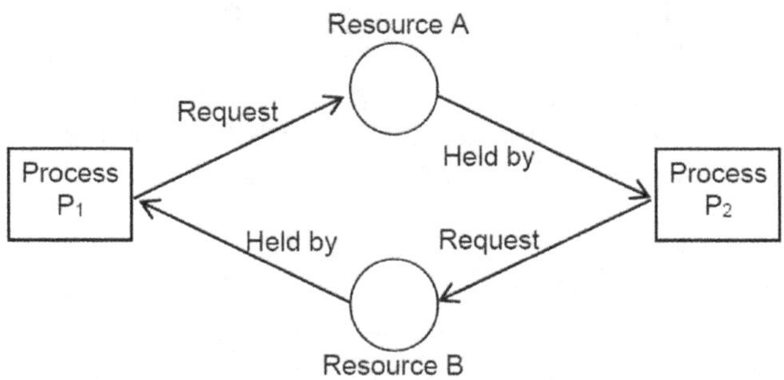

Figure 1.11a: Deadlock Scenario

A is allocated to P_2 and B is allocated to P_1. If P_1 request A and P_2 request B, then there will be a deadlock because A will be released only when P_2 gets B and B will be released only when P_1 gets A.

1.7.2 Deadlock Characterization

A deadlock situation can arise if the following four conditions hold simultaneously in a system.

- **Mutual Exclusion:** Shared resources are acquired and sued in mutual exclusive manner, that is, by at most one process at a time. If another process request that resource, it must be delayed until it has been released.
- **Hold and Wait:** A process must be holding at least one resource and waiting to acquire additional resource that are currently being held by other process.
- **No Preemption:** Resources can not be preempted means resources previously granted can not be forcibly taken away from a process. They must be explicitly released by the process holding them.

Circular Wait: There is a circular chain of two or more processes, each of which is waiting for a resources held by the next member of the chain.

All of these four conditions must be present for a deadlock to occur. If one of them is absent, no deadlock is possible.

Deadlock vs Starvation

A deadlock occurs when a set of processes in the system is blocked waiting on requirements that can never be satisfied. While starvation occurs when a process waits for a resource that continually becomes available but is not assigned to that process because of priority or flaw in design of the scheduler.

In starvation, it is not certain when the process will get the requested resource, but in deadlock, it is certain that process will never get the requested resource

In starvation, the utilization of resource under contention is 100 % while in deadlock, it is 0 %.

1.7.3 Methods of Handling Deadlock

Principally, we can deal with deadlock problem in one of the three ways:

- Deadlock Prevention
- Deadlock Avoidance
- Deadlock detection & Recovery

Deadlock Prevention

1. The basic philosophy of deadlock prevention is to deny at least one of the four necessary conditions for deadlock. Let us elaborate on this approach by examining each of four necessary conditions.

- **Mutual Exclusion:** In general, mutual exclusion for the resources cannot be disallowed because resources may be either sharable or non-sharable. For example, a printer cannot be simultaneously shared by several processes. On the other hand, sharable resources do not require mutually exclusive access. A process never needs to wait for sharable resource. For example read only files are sharable resources. If several processes attempt to open a read only file at the same time, they can be granted simultaneous access to the file.

  Thus mutual exclusion must be supported by OS if there are non-sharable resources in system.
- **Hold and Wait:** To ensure that the hold and wait condition never occurs in the system, we must guarantee that, whenever a process requests a resource, it does not hold any other resources.

There are two approaches to prevent hold and wait condition:

- The process request all needed resources prior to commencement of execution. This strategy requires that all of the resources, a process will need, must be requested at once. The system must grant them on an "all or none" basis. If the set of resource needed by a process is available, then system may grant all of those resources to the process at one time and process will be allowed to proceed. If the complete set up resources needed by a process is not currently available, the process must wait until complete set is available.

- The process requests resources incrementally in the course of execution but releases all its resource holding upon encountering a denial.

Note: First approach suffers following disadvantages:

1. **Low Resource Utilization:** Since many resources may be allocated but remains unused for long time.
2. **Starvation:** A process that needs several popular resources may have to wait indefinitely, because at least one of the resources that it needs is unavailable
3. **No Preemption:** The no-preemption condition can be denied by allowing preemption, that is authorizing the system to revoke ownership of certain resource from blocked process. For this there are following approaches:
    - If a process is holding some resources and requests another resource that cannot be immediately allocated to it, then all the resources currently being held are preempted. The preempted resources are added to the list of resources for which process is waiting. The process will be restarted only when it can regain its old resources as well as new ones that it is requesting.
    - If a process requests, we first check whether they are available. If they are available allocate them. If they are not available, check whether they are allocated to some other process that is waiting for additional resources. If so, preempt the desired resources from waiting process and allocate them to requesting process. If the resources are not either available or held by a waiting process, requesting process must wait.
4. **Circular Wait:** The circular wait condition can be prevented by defining a linear ordering of resource types and granting the resources in increasing order of enumeration.

In this approach all the resources are given a unique number in increasing order. Suppose $R = \{R_1, R_2, ..R_m\}$

If a process acquires resource R_j then it can only request resources R_{j+1} or higher thereafter. Thus circular wait condition can be prevented as shown in figure 1.11b.

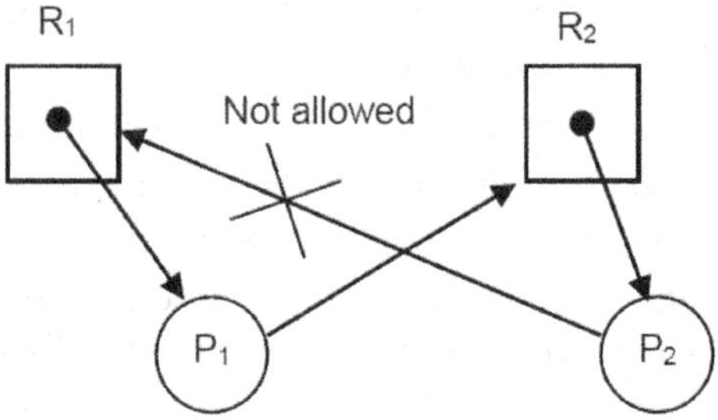

Figure 1.11b: Prevention of circular wait

Deadlock Avoidance

In the deadlock prevention, we constrain resource requests to prevent at least one of the four conditions of deadlock from occurring. This leads to:

- Low resource utilization
- Reduced system throughput
- Starvation

The basic idea of deadlock avoidance is to grant only those requests for available resources that cannot possibly result in a state of deadlock. This strategy is usually implemented by having the resource allocator examine the effect of granting a particular request. If the granting of resource cannot possibly lead to deadlock, the resource is granted to requestor otherwise requesting process is suspended until such time when its pending request can be safety granted. Thus, this approach requites knowledge of future requests of resources.

Deadlock Detection and Recovery

In this approach, resources are granted to requesting processes without any check. Periodically system state is examined to determine if a set of processes is deadlocked. Whenever deadlock is found, it is recovered by either aborting all the deadlocked process or by aborting a victim process. The victim process is selected by following considerations:

- How much time the process will take to complete.
- From how much time process is being executed.
- How many resources it has acquired.
- How many resources it will need to complete.
- What is the priority of process
- etc.

While selecting the victim the fact must be in mind that any process should not be selected as victim repeatedly.

1.7.4 Models of Deadlock

Depending on the type of requests of process, there are four types of deadlocks:

1. **Single-Unit Request Model:** In this model a process is restricted to requesting only one unit of resource at a time (i.e., request the resource one by one). Thus out degree of nodes in WFG is one. Therefore, if all the resources have single instance, a cycle in deadlock depicts deadlock.
2. **AND Request Model:** In this model a process can request multiple resources simultaneously, and it remains blocked until it is granted all of the resources. The AND request model is more powerful and allows more concurrency as a process can request several resources simultaneously.

   Note:

   - Out degree of nodes in Wait For Graph can be greater than one in this model.
   - A process can be involved in more than one deadlock in this model while in single-unit request model a process can be involved in only one deadlock.
   - Provided that there is only one instance of every resource, a cycle in Wait For Graph depicts deadlock.
3. **OR Request Model:** In the OR request model, a process can simultaneously request multiple resources and it remains blocked until it is granted any one of requested resources.

Note:

- A cycle in the wait for graph is not a sufficient condition for a deadlock in this model even if there is only one copy of each resource in the system.

- A typical example of OR request model is read request for a replicated data object, where reading any copy of the data object satisfies the request. Suppose R_2 and R_3 is replicated data object to which processes P_2 and P_3 is using respectively (Figure 1.12).

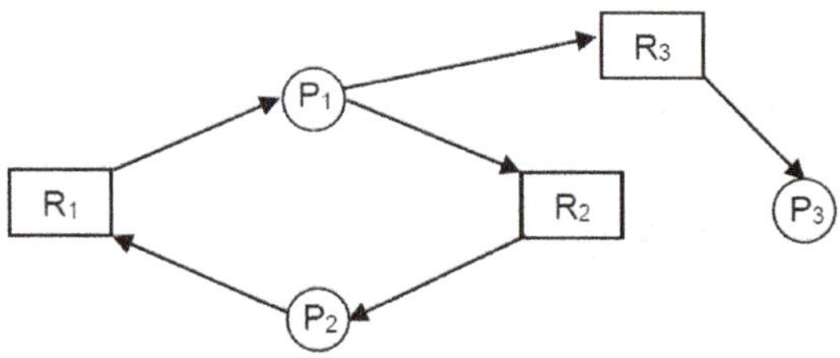

Figure 1.12: Resource Allocation Graph Example

- Although there is a cycle but there is no deadlock because as soon as P_3 releases R_3, P_1 will get it and will remove request for R_2 so cycle is removed.

Wait For Graph for above Resource Allocation Graph is shown Figure 1.13

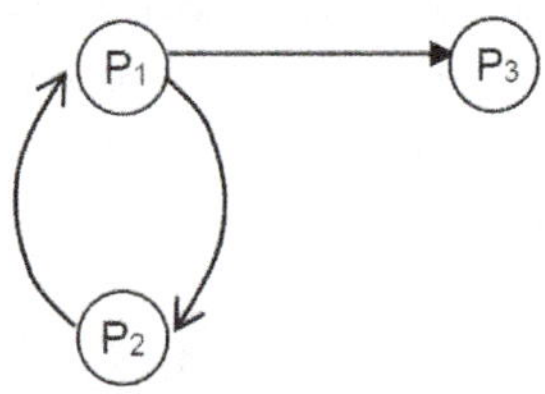

Figure 1.13: Wait For Graph Example

Note:

- The presence of cycle in Wait For Graph is not a sufficient condition for deadlock even if there is only one instance of every resource. This is just because if a process is involved in a cycle, it means that process may be unable to get only one of the several requested resource. The process can still get one of requested resources which do not lie on the cycle.
- The presence of knot in Wait For Graph is sufficient condition for deadlock.

4. **The AND-OR Request Model:**
    - Generalization of AND & OR model.

- Knot is a sufficient condition for deadlock.
- Process requests are specified by using a predicate whose variables are resources.
- For example request "R_1 AND (R_2 OR R_3)" can be satisfied by either "R_1 AND R_2" or "R_1 AND R_3"

5. **The P-Out of Q Request Model:**
   - A process simultaneously requests Q resources and remains blocked until it is granted any P of those resources.
   - A knot is sufficient condition for deadlock.

Note:

- Single request model is a special case of AND request model.
- Single request model is also a special case of OR request model.
- AND & OR request models are special cases of P-out of-Q request model. When P = Q it is AND request model while when P = 1 it is OR request model.

1.7.5 Types of Resources

Reusable Resources:

- Are not consumed by use.
- Has fixed number of units of resource.
- These units of resource can neither be created nor destroyed.
- When allocated to process, they are held until the process is done and then released.
- Examples: CPU, Memory, I/O devices.

Consumable Resources:

- They vanish as a result of their use.
- When allocated to a process, they are consumed and cease to exit.
- There are one or more producers of resource.
- No fixed number of units of resource, since resource units can be created and consumed.
- Producer, if not blocked, can produce any number of units of resource.

- Examples: messages, interrupt signals, V operation in semaphore.

Types of Resource Access:

Resource can be accessed in two modes:

- **Exclusive:** Two or more processes cannot access a resource simultaneously in exclusive access mode.
- **Shared:** A resource can be accessed by any number of processes simultaneously.

1.7.6 General Resource System

General resources system is characterized by the following:

1. A non empty set of processes $P = \{P_1, P_2, \ldots\ldots P_n\}$.
2. A non empty set of resources $R = \{R_1, R_2 \ldots\ldots R_n\}$.
   - R can be partitioned into two disjoint sets:
     - R_R – a set of reusable resources
     - R_C – a set of consumable resources
   - For each reusable resource, there exists a non-negative number denoting the total number of units of resource present in the system.
   - For each consumable resource, there exists a set of processes, that can produce that resource (known as producer).

Graphical Representation of General Resource System:

- Wait-For-Graph
- Resource Allocation Graph
- General Resource Graph

Wait For Graph: WFG is characterized by:

- It is a directed graph.
- A set of process $P = \{P_1, P_2, P_3 \ldots\ldots P_n\}$
- An edge from P_i to P_j ($P_i \rightarrow P_j$) denotes "Process P_i is waiting for P_j to release the resource that P_i needs.

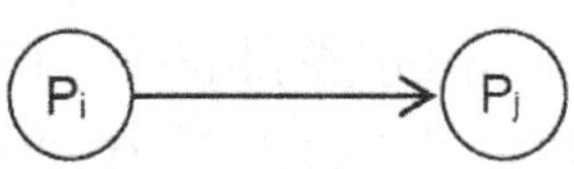

(P_j needs resource held by P_j)

Note: Some graph related specifications :

- A node y is **reachable** from node x only if there exists a path from node x to node y.
- A **cycle** in a graph is a path that starts and ends on same node.
- A node in a directed graph is **sink** only if it has incoming edges to it.
- A **knot** in a graph is a subset K of nodes such that the reachable set of each node in K is exactly K.

Knot examples are shown in figure 1.14.

Examples:

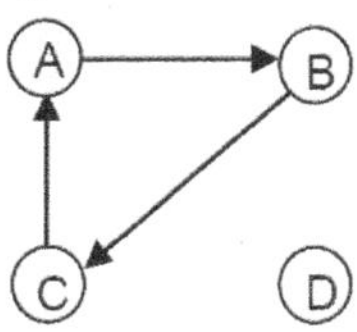

(A, B, C are in cycle also in not)

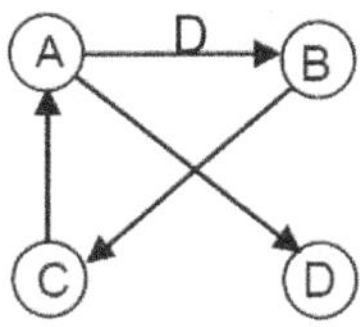

- (A, B, C are in Cycle but not in a knot)
- D is sink

Figure 1.14: Knot examples

1.7.8 Graphical Representation of General Resource System

Resource Allocation Graph

The resources are partitioned into several types, each of which consists of some number of identical instances. If the system has two CPUs then the resource type CPU has two instances. Similarly resource type printer may have five instances.

If a process request an instance of resource type, allocation of any instance of the type will satisfy the request.

The resource allocation graph is used to determine deadlock in a system.

This graph consists of set of vertices V and set of edges E. The set of vertices V is partitioned into two different types of nodes as:

- Set of all active processes in the system
- Set of all resource types in the system.

Then

Resource Allocation Graph = {V, E}

V = {P, R}

$P = \{P_1, P_2, P_3, P_4 ....\}$

$R = \{R_1, R_2, R_3 ......\}$

Edges

$P_i \rightarrow R_j$	Signifies that process P_i requested an instance of resource type R_j and is currently waiting for that resource. It is called request edge.
$R_j \rightarrow P_i$	Signifies that an instance of resource type R_j ahs been allocated to process P_i. It is called assignment edge.

In Resource Allocation Graph, process is represented as circle and each resource type as square. Since resource type may have more than one instance, such instances are represented as dot with in square.

Rules

When a process P_i requests an instance of resource type R_j, a request edge is inserted in Resource Allocation Graph. When this request can be fulfilled, the request edge is instantaneously transformed to assignment edge. When the process no longer needs access to the resource, it releases the resource and as a result assignment edge is deleted.

Conditions of Deadlock in Resource Allocation Graph

- If each resource type has exactly one instance, then a cycle in Resource Allocation Graph implies that deadlock has occurred.
- If each resource type has several instances, then a cycle does not necessarily imply that a deadlock has occurred. In this case, a cycle in the graph is necessary but not sufficient condition for existence of deadlock.

Consider the following set of P, R and E.

$P = \{P_1, P_2, P_3,\}$

$R = \{R_1, R_2, R_3, R_4\}$

$E = \{P_1 \rightarrow R_1, P_2 \rightarrow R_3, R_1 \rightarrow P_2, R_2 \rightarrow P_2, R_2 \rightarrow P_1, R_3 \rightarrow P_3\}$

Resource Instances:

- R_1 has one instances
- R_2 has two instances
- R_3 has one instances
- R_4 has three instances

Resource Allocation Graph of given scenario is shown in figure 1.15.

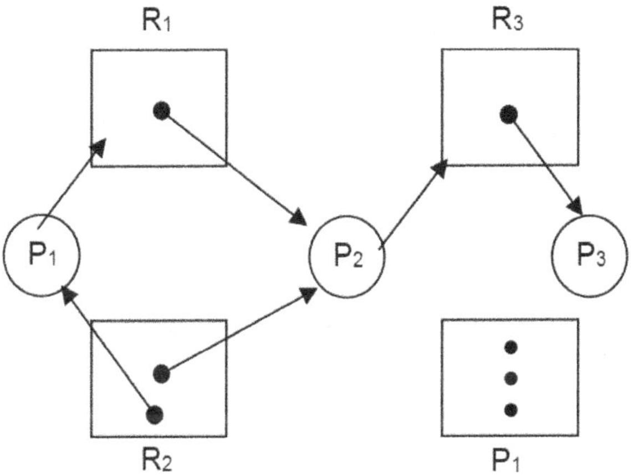

Figure 1.15: Resource Allocation Graph

There is no cycle in the graph so no process in the system is deadlocked.

Suppose process P_3 requests an instance of resource type R_2. Since no instance of R_2 is available, $P_3 \rightarrow R_2$ is added to graph. At this point two cycles exist in the system.

$P_1 \rightarrow R_1 \rightarrow P_2 \rightarrow R_3 \rightarrow P_3 \rightarrow R_2 \rightarrow P_1$

$P_2 \rightarrow R_3 \rightarrow P_3 \rightarrow R_2 \rightarrow P_2$

Here P_1, P_2 and P_3 are deadlocked because P_2 is waiting for the resource R_3 which is held by P_3. On the other hand P_3 is also waiting for either for P_1 or P_2 to release R_2. In addition P_1 is waiting for process P_2 to release R_1 (Figure 1.16).

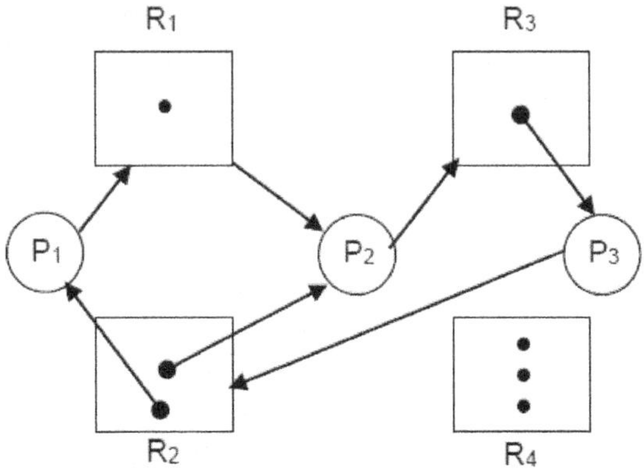

Figure 1.16: Updated Resource Allocation Graph

Consider another Resource Allocation Graph as shown in **figure 1.17:**

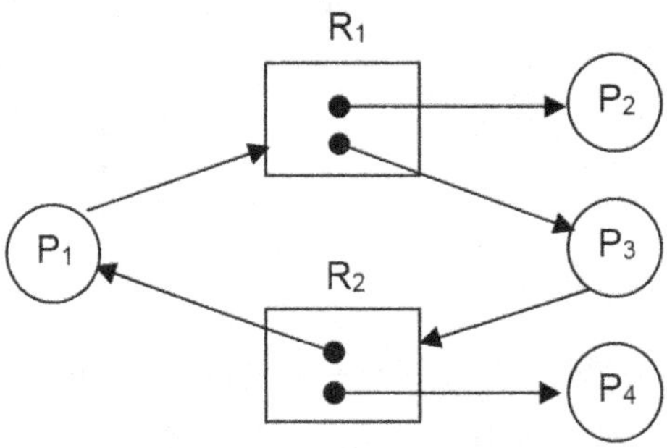

Figure 1.17: Resource Allocation Graph Example

In this example there is a cycle:

$P_1 \rightarrow R_1 \rightarrow P_3 \rightarrow R_2 \rightarrow P_1$

But there is no deadlock because P_4 may release its instance of resource R_2 which can then be allocated to P_3, breaking the cycle.

General Resources Graph

A general resource graph is bi-partite directed graph (whose nodes can be divided into two disjoint sets such that two adjacent nodes in graph can not come from same set).

Therefore the components of GRG are:

1. **Vertices of General Resources Graph:**

   P-Set of Processes P_1 ...... P_n

   R-Set of Resources R_1...... R_m

   - R can be divided into two disjoint sets of consumable and reusable resources.
   - For every reusable resources, a non negative number t_i denotes total no of R_i

2. **Edges of General Resources Graph**

   - **Request:** Directed from P to R
   - **Assignment:** Directed from reusable R to P
   - **Producer:** Directed from consumable R to P.

Note:

- Request edge indicates that process is requesting one unit of resource.
- Assignment edge indicates that process has been assigned one unit of that resource.
- Producer edge indicates that the process is a producer of that resource.

Available Vector: A non-negative integer vector (r_1, r_2, ... r_n) called available units vector, denotes the no of units of the resource available in any state.

A general resource graph must satisfy the following conditions:

1. For every reusable resources:

   - Total assigned units of R_i < = initial number of units of R_i (t_i)
   - Available units of R_j (r_j) = initial units of R_i (t_i) – total assigned units of R_i
   - For each process P_j

     Assigned units of R_i to P_j + Requested units of R_i by P_j < = total number of units of resource R_i (t_i)

     That is, at any instant, a process cannot request more than total units of reusable resource.

2. For every consumable resource:

   - There is an edge from R_i to P_j only if P_j is a producer of R_i
   - Available units (r_j) > = 0.

In GRG, a process is denoted by rectangle and a resource by circle. A consumable resource is denoted by thick circle. The little circle inside a resource denotes the resource units.

Process Operations and Effects on the General Resource Graph

The system state and the corresponding general resource graph change as the process execute operations. The process can perform the following operations:

- Request
- Acquisition
- Release

These operations affect the system state in following ways:

1. A request operation results in addition of request edge ($P_i \rightarrow R_j$) in General Resource Graph.
2. If e units of R_i is acquired by P_j then
   - Available unit r_i is decreased by e.
   - For reusable resources, e request edges are replaced by assignment edges.
   - For consumable resource, all e request edges are deleted.
3. Release occurs when process does not need R_j any more. When a process P_i releases e units of resources e assignment edges (R_j to P_i) in General Resource Graph are deleted and available unit r_j of R_j is incremented by e.

Example: A symbol of a system may be as:

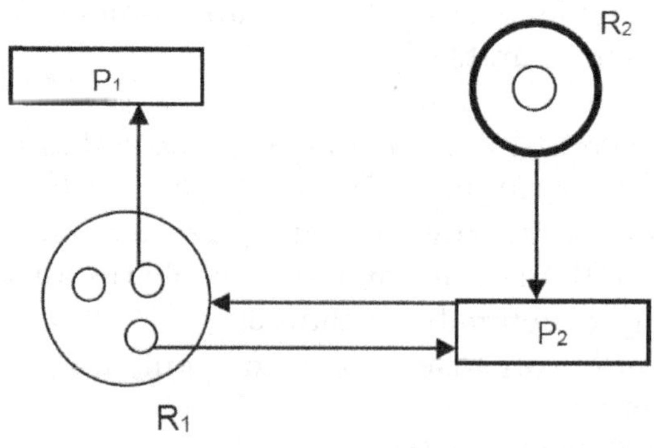

Here:

- R_2 is consumable resource.
- P_2 is producer of R_2
- P_1 and P_2 both are holding one unit of R_1 and P_2 is requesting one more unit of R_1

If P_1 makes request for 2 units of R_1 one unit of R_2

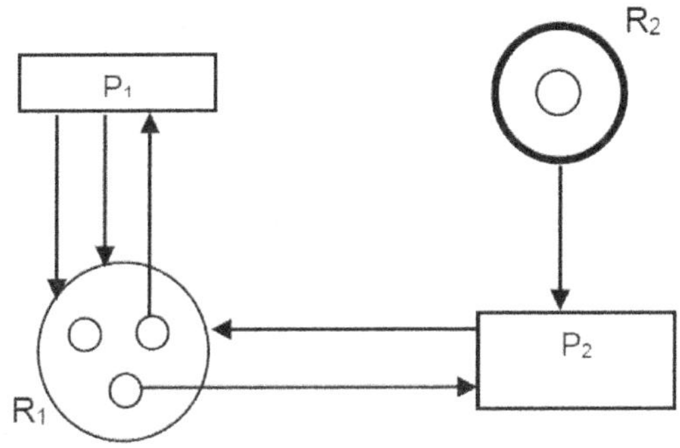

P_1 acquires one unit of R_2

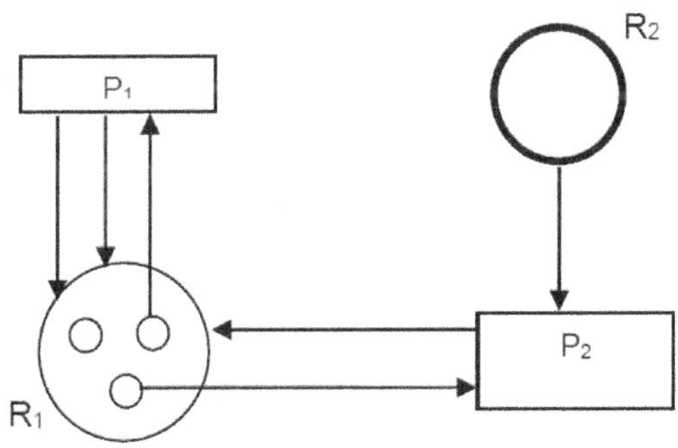

P_2 acquires one unit of R_1

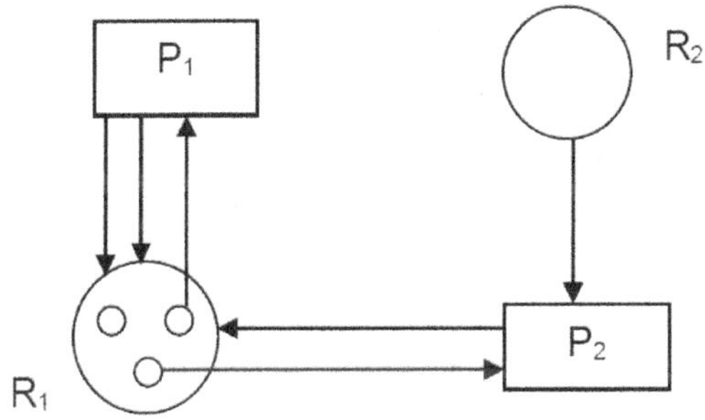

P_2 produces three units of R_2 and releases one unit of R_1

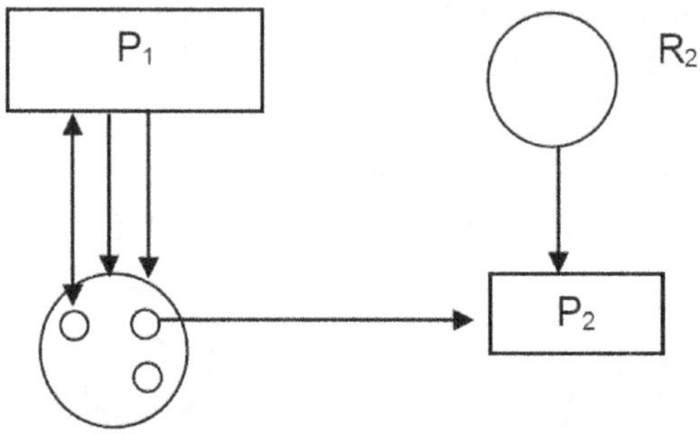

1.7.9 Deadlock Analysis Using General Resource Graph

- A process is blocked iff for some resource (R_i), the number of requested units is more than available units (r_i) of that resource.
- A blocked process will lead to a deadlock if there is no way for it to become unblocked. A process is not deadlocked if and only if we can find a sequence of operations that leaves the process unblocked.
- A General Resource Graph is a powerful theoretical tool – it gives the simplified view of the state of system.

Theorems linking graph properties to presence / absence of a deadlock are as follows:

Graph Reduction Method: A reduction is most optimistic set of operations that unblocks one or more blocked processes.

Reducing general resource graph by an unblocked process P_i.

For each reusable resources (R_j)

- Delete all assignment and request edges.
- For each assignment for a reusable resource, increment the number of available units (r_j)

For each consumable resource (R_j)

- Decrement the available unit (r_j) by number of request edges.
- Delete all request and producer edges.
- It P_i is producer of R_j, then set available unit (r_j) for that resource to ∞.

After a general resource graph has been reduced by a process, some additional resources are freed. Consequently more processes become unblocked, which can then be used in next reduction and so on. Thus reduction can be repeatedly performed on General Resource Graph. At any stage of reduction, there may be several processes that can be reduced, so any one can be selected randomly for reduction. The final outcome can be one of the following:

- All processes are reduced.
- The graph reaches a stage where no further process reduction is possible.

Deadlock Determination by Using General Resource Graph

We can classify the approaches of Deadlock Determination by Using General Resource Graph in three ways:

- Good Approach
- Better Approach
- Best Approach

1. Good Approach:

Note:

- A general resource graph is **completely reducible** if a sequence of reduction deletes all edges in the graph.
- A system state is **deadlock free** if its general resource graph is completely reducible.

Theorem: A process P_i is not deadlocked in General Resource Graph if and only if a sequence of reductions applied to that graph leaves the graph in a state in which P_i is unlocked.

Thus, the reducibility of a General Resource Graph implies that corresponding state is deadlock free. However, the reverse is not true. That is, if a system state is deadlock free, its General Resource Graph need not to be completely reducible. Therefore, lack of reducibility of General Resource Graph does not imply that corresponding state is deadlocked.

Therefore, the last theorem and note (b) state the sufficient condition for a process and system state to be deadlock free, but they do not give a sufficient method for detecting deadlock.

2. Better Approach

Expedient State: "A state is an expedient state if all processes having outstanding requests are blocked".

Note:

- The blocked means waiting to be assigned a resource when sufficient units of that resource are not available. Thus, a state where in a process is waiting for assignment of some units of resources and at least that many units of that resources are currently available, then the state is not expedient (Figure 1.18).

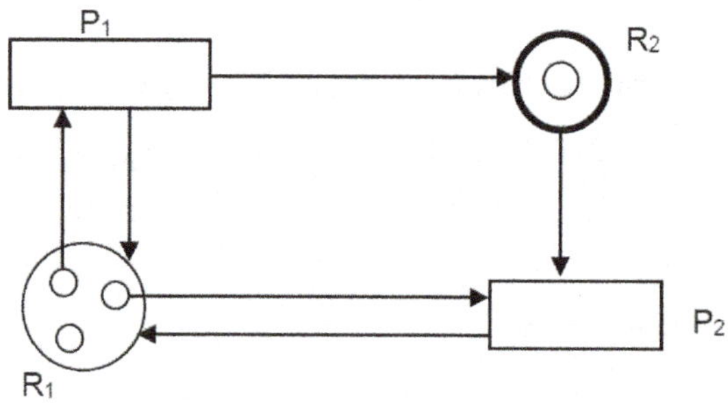

Figure 1.18: Non-Expedient

In the **figure 1.18**, state is not expedient because both P_1 and P_2 are waiting in this state even through the request of one can be met.

A state that is not expedient becomes expedient only if all possible grantable resources have been granted (Figure 1.19).

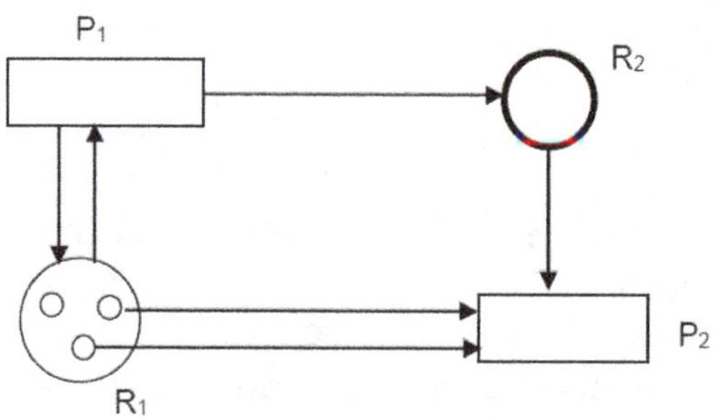

Figure 1.19: Expedient

Theorem: In a general resource graph:

- A cycle is necessary condition for a deadlock.
- If a graph is expedient, then a knot is a sufficient condition for deadlock.

Thus, a deadlock in system state implies the cycle in corresponding General Resource Graph. Also if the graph is expedient, the existence of knot implies deadlock. But the absence of knot dose not implies absence of deadlock.

So in the absence of knot in General Resource Graph, these may be deadlock there may not be (Figure 1.20)

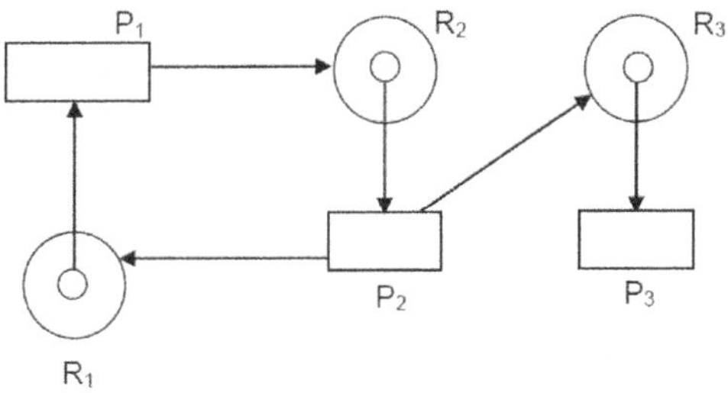

Figure 1.20: Deadlock with no Knot

Note:

- No node in the knot is sink or has a path to a sink.

Theorem: If in a expedient General Resource Graph a node P_i is not a sink and no path starting from P_i, leads to a sink then P_i is deadlocked.

As the absence of Knot in General Resource Graph does not imply the freedom from deadlock, so this approach is not efficient.

Best Approach

For Single Unit Systems

Consider a general resource graph G for

- A system where processes can request one unit of resource at a time

  Theorem: If the G is expedient then knot is sufficient and necessary condition for deadlock.
- A system with single unit reusable resources only

Theorem: If G is expedient then cycle is necessary and sufficient condition for deadlock.

For Systems with only Reusable Resources

Theorem: The system is deadlock free only if its corresponding General Resource Graph is completely reducible.

Systems with only Consumable Resources

A consumable resource system is characterized by claim-limited graph and this graph is used in deadlock determination.

Claim-Limited Graph: Claim limited graph of a consumable resource system represents the worst-case situation of a system where all the units of all the resources have been exhausted and every consumer of every resource is requesting one unit of all resources that it could possibly request.

The Claim limited graph for a system is constructed by:

- Making each consumable resource have zero available units.
- Having a request edge (P_i, R_j) iff P_i is consumer of R_j.

Example: Claim limited graph of a system that has two consumable resources R_1 and R_2 and two process P_1 & P_2. P_1 is producer of R_1 and is a consumer of R_2. P_2 is producer is R_2 and is a consumer of R_1

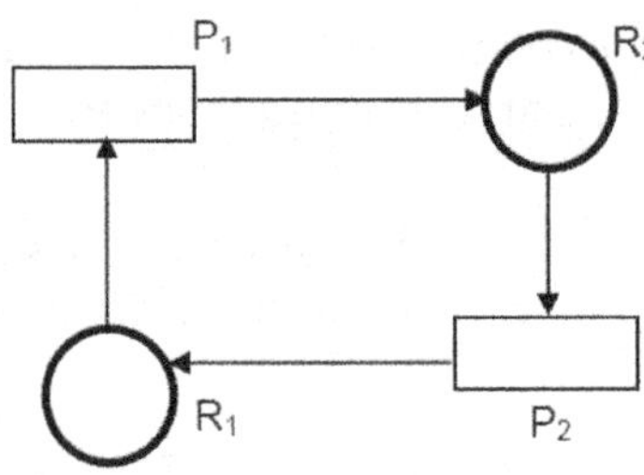

Theorem: A consumable resource only system is deadlock free if its claim-limited graph is completely reducible.

As in above claim-limited graph there is no unblocked process so it cannot be reduced. Thus the system is not deadlock free.

If in the above example, system has a third process that was producer of any of these resources and was not consumer of any one, then the system would have been deadlock free.

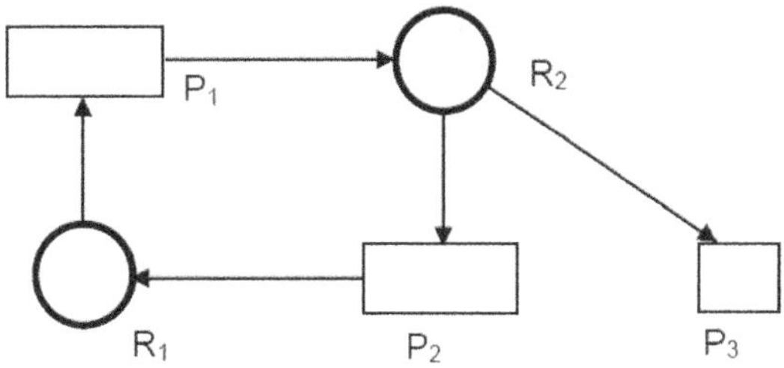

Note: The deadlock freedom in consumable resources systems only, requires that there be at least one producer of some resources that is not a consumer of any resources.

Table of Deadlock Theorems

	Deadlock Freedom Conditions	**Deadlock Presence Conditions**
General Cases	CR ⇒ DLF	DL ⇒ CY Kn, Exp ⇒ DL
Reusable resources only	CR ⇒ DLF	
Reusable + single unit resource		Cy; Exp ⇔ DL
Single unit resources		Kn, Exp ⇔ DL
Consumable Resources	CR Claim-limited Graph ⇒ DLF	

CR: Completely Reducible

DLF: Deadlock Free

DL: Deadlock

CY: GRG Contains Cycle

KN: GRG Contains Knot

Exp: GRG is expedient

EXERCISE

Q1. What do you understand by Operating system? What are its main functions?

Q2. Explain the layered approach of Operating System. What are the advantages of layered approach in system design?

Q3. Show how operating system works as resource manager.

Q4. Discuss the various criteria that led to development of new operating system.

Q5. Explain how multiprogramming contributes towards higher CPU utilization and increased throughput.

Q6. Distinguish between multiprogramming and multiprocessing. What were the key motivations for development of each?

Q7. State why SPOOLing is necessary in batch operating system? Also explain how it is beneficial in multiprogramming environment.

Q8. Can a uniprocessor system have no ready process and no running process? Is this a "dead" system? Explain your answer.

Q9. Differentiate followings:

a. Distributed and network operating system

b. Symmetric and asymmetric multiprocessing

Q10. What is a process? Explain various states of process through sate transition diagram. Illustrate the mechanism for interaction between two concurrent processes.

Q11. What do you mean by process synchronization? Dining Philosopher Problem is one of the most beautiful problems of process synchronization. Discuss what it is? How it can be solved?

Q12. Briefly describe the operating tasks involved in Context Switching.

Q13. Give the several reasons why the concurrency factor is appropriate and important for design of operating systems.

Q14. What is Producer/Consumer problem in context of concurrent processes? Also give the semaphore solution for it.

Q15. Why might the following be unacceptable?

```
Cobegin
a= b + c
d=b * c - x
e=a / 6 +n ** 2
coend
```

Q16. Explain the path expression approach for process synchronization. Explain what the following path expressions do:

```
Path {read} + write end
Path {open + read}; close end
```

Q17. Discuss about various models of deadlock.

Q18. What are the different types of resources present in the system

Q19. What do you understand by general resource graph. How it is helpful for resource management in distributed system.

Q20. Discuss the necessary and sufficient condition for deadlock. Discuss the graph reduction method to test a system state for deadlock. Prove the existence of a cycle is necessary condition for occurrence of deadlock.

Q21. Write short note on following:

1. Expedient State
2. Claim Limited Graph
3. Knot

Q22. Draw the claim-limited graph of the consumable resources only system that has two consumable resources R1 and R2 and three processes P1, P2 and P3. P1 is the producer of R1 and is a consumer of R2. P2 is the producer of R2 and is a consumer of R1. P3 is the only producer of R1. Determine whether the system is deadlock free or not. Give reasons also for your answer.

Q23. Discuss the necessary and sufficient conditions for deadlock in the following systems:

1. Systems with single unit resources
2. Systems with only reusable resources
3. Systems with Consumable resources

2

Distributed Systems

2.1 DISTRIBUTED SYSTEMS

2.1.1 Introduction

Distributed system is a collection of independent computers that appears to its users as single coherent systems.

Distributed computer systems are characterized by:

- Collection of autonomous computers capable of communication via their hardware and software interconnections.
- Absence of shared memory and common clock.
- Computers has it's own memory and runs it's own OS.

In other words, distributed system is a collection of loosely coupled processors interconnected by communication network. (On the contrary, multiprocessing system is a collection of tightly coupled processors, which share the common memory and clock).

Each computer in distributed system owns its own resources and has the ability to access the resources of other machine over the network (Figure 2.1 and 2.2). The resources owned and controlled by a computer are said **local** to it, while the resources owned and controlled by other computers and those that can only be accessed through network are said **remote** resources.

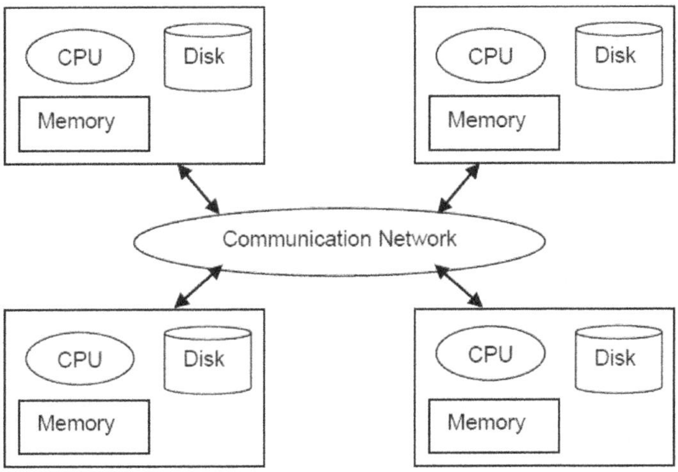

Figure 2.1: Distributed System

A distributed system is organized as middle ware. Note that middle ware layer extends over multiple machines.

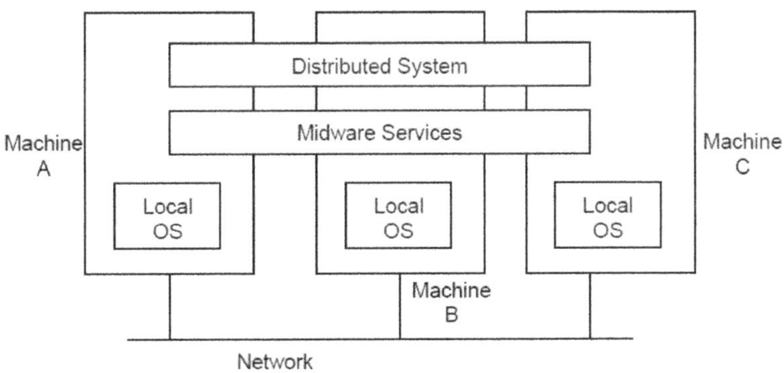

Figure 2.2: A distributed as middle ware which extends over multiple machines

2.1.2 Distributed Operating System

A distributed OS governs the operation of distributed system and provides a virtual machine abstraction system to its user.

It looks to its user like an ordinary centralized operating system but provides the user with transparent access to the resources of number of machines.

The key concept of distributed system is transparency. The use of multiple processors and accessing of remote data is invisible (transparent) to its user. The user views the system as a virtual uniprocessor and not as a collection of distinct machines – for instance, a user simply submits a job to the distributed OS through a computer. Distributed OS performs distributed execution of job. The user does not know on what computers the job was executed, on what computers the files needed for execution was stored or how communication or synchronization on different computers were carried out.

2.1.3 Network Operating System

Network OS is different from distributed OS.

It provides an environment in which users, who are aware of multiplicity of machines, can access remote resources by either logging into appropriate remote machine or transferring data from remote machine to own machine (Figure 2.3).

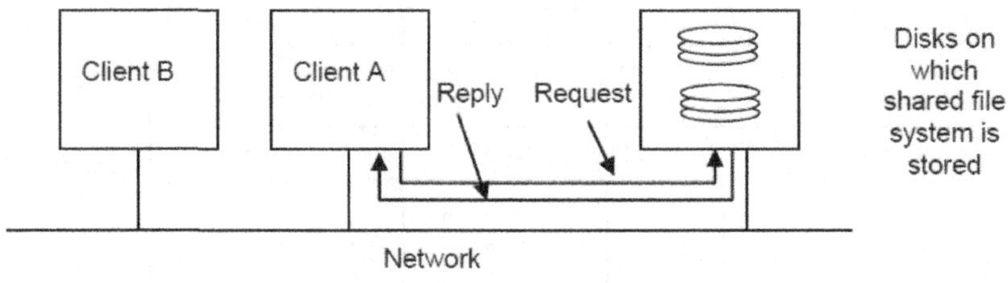

Figure 2.3: Network system

The network OS provides some tools for logging and remote file transfer (ftp, telnet etc).

2.1.4 Advantages of Distributed System

- Transparency
- Resource sharing
- Computation speed up
- Reliability
- Availability
- Fault tolerance

- Information sharing
- Modular expandability

Transparency implies the ability to access a resource or service without having to know its location in system.

Transparency	Description
Access	Hide differences in data representation and how a resource is accessed
Location	Hide where a resource is located
Migration	Hide that a resource may move to another location
Relocation	Hide that a resource may be moved to another location while in use
Replication	Hide that a resource may be shared by several competitive users
Concurrency	Hide that a resource may be shared by several competitive users
Failure	Hide the failure and recovery of a resource
Persistence	Hide whether a (software) resource is in memory or on disk

If a number of different nodes with different capabilities are connected to one another, then user at one site may be able to use resources available at other site.

If a particular computation can be partitioned into sub-computations that can run concurrently, then distributed system allows us to distribute sub-computations among various sites and they can be run concurrently. Thus the computations are speed up.

In addition, if a particular site is currently overloaded with jobs, some of them may be moved to other lightly loaded sites. This movement of job is known as load sharing, which also speeds up computation.

A distributed computing system provides improved reliability and availability because a few components of the system can fail without affecting the availability of the rest of the system. Also through replication of data and services, distributed systems can be made fault tolerant.

Information Sharing: When several sites are connected to one another by a communication network, the users at different sites have the opportunity to exchange information.

Note: Load balancing can be performed in two ways:

- Initiate the processes of a computation in different nodes of the system.
- By moving some processes of overloaded node to another lightly loaded node. This is called **process migration**.

2.1.5 Important Issues in Distributed Operating System

Some important issues that arise in the design of distributed OS include the:

- Unavailability of up-to-date global knowledge
- Naming
- Scalability
- Compatibility
- Process synchronization
- Resource management
- Security
- Openness

1. Global Knowledge: In the case of uniprocessor-centralized system, the up-to-date state of all processes and resources (global state of system) is completely and accurately known.

In the distributed systems, this issue takes on new dimensions and its solution becomes more complex for the following reasons:

- Due to unavailability of global memory and global clock.
- Due to unpredictable message delay.

Thus it is practically impossible for a computer to collect up-to-date information about the global state of distributed computing system. Therefore, a fundamental problem in the design of distributed operating system is to determine efficient techniques to implement decentralized system wise control, where a computer does not know the current and complete status of the global state.

2. Naming: The names are used to refer the object. The objects that can be named in computer system may be, computers, printers, services files and users.

Name resolution refers to the process of mapping a name to an object. A **name space** is collection of names, which may or may not share an identical resolution mechanism.

Name resolution is done by service named as **name service**, which maps the logical name into physical address by using the table lookup. In the implementation of table lookup, tables (also known as directories) that store names and their physical addresses are used for mapping names to their addresses.

3. Scalability: The issue of scalability deals with the suitability of the design of a system to cater the demands of growing system. The system generally grows with time. The techniques used in designing a system should not result in system unavailability or degraded performance when growth occurs.

For example, broadcast based protocol work well for small number of systems but not for large number of systems. Consider a distributed file system that locates files by broadcasting queries. Under this file system, every computer in distributed system is subject to message handling overhead, irrespective whether it has requested file or not. As the number of users increase, the number of file location queries will increase and overhead will grow larger, hurting the performance of every computer.

Therefore in designing the distributed system, scalability must be in mind, and if in the long run growth occurs, it should be not be at cost of performance.

Distributed systems operate effectively and efficiently at many different scales, ranging from small intranet to Internet. A system is described as scalable, if will remain effective when there is a significant increase in number of resources and number of users.

4. Compatibility: Compatibility refers to interoperability among resources in a system. The three different levels of compatibility exist in distributed systems.

- Binary level
- Execution level
- Protocol level

In binary level compatibility, all processors execute the same binary instruction set, even though the processors may differ in performance. This type of compatibility is useful for system development as the code for many functions, provided by system programs directly depend on under laying machine level instructions.

Note: Distributed systems under binary level compatibility cannot include computers with different architectures from the same or different vendors.

Execution level compatibility is said to exist in distributed systems if the same source code can be compiled and executed properly on any computer in the system.

Protocol level compatibility achieves interoperability by requiring all system components to support a common set of protocols.

Note: A significant advantage of this compatibility is that individual computers can run different operating systems while not sacrificing their interoperability.

5. Openness: The openness of computer system is characteristic that determines whether the system can be extended or not. The openness of distributed system is determined by degree to which new resource sharing services can be added and made available for use.

6. Process Synchronization: In centralized system where all the processes reside in same memory, so it is easy to enforce mutual exclusion for synchronizing them. But in distributed system in the absence of shared memory, it is difficult because process running at different computers may try to access shared resource. Therefore concurrent access to shared resource by several uncoordinated user requests must be serialized to secure integrity of shared resource.

7. Resource Management: The resource management in distributed system is concerned with making both local and remote resource in same manner. In other words, the location of resources should be hidden from user.

The resources of distributed system are made available to users in following ways:

- Data Migration
- Computation Migration
- Process Migration

a. Data Migration: In the process of data migration, data is brought to the location of computation that needs access to it, by distributed operated system.

Suppose that a user on site A wants to access data (such as file) that resides at site B. The data migration can take place in one of two basic methods:

1. Transfer the entire file to site A. Thus all access of site A to remote file of site B becomes local. A copy of file (if it has been modified) is sent back to site B. *Disadvantage:* Even if only a modest change has been made to a large file, all the data must be transferred from B to A and A to B as well.
2. The second approach is to transfer to site A only those portions of the file that are actually necessary for immediate task. If another portion is required later, another transfer will take place. When the user no longer wants to access file any part of it that has been modified must be sent back to site B.

Note: If only a small part of the large file is being accessed, the latter approach is preferable. If significant portion of the file are being accused, it is more efficient to copy the entire file.

b. Computation Migration: In some circumstances we may want to transfer computation, rather than the data, across the system. This approach is called computation migration.

For example, consider a job that needs to access various large files that reside at different sites, to obtain summary of those files. It would be more efficient to access the files at sites where they reside and return the desired results to the site that initiated the computation.

Note: If the time to transfer the data is longer than the time to execute remote command, it would be better to migrate computation rather than data.

The migration of computation can be carried out in two ways —

- By RPC mechanism
- By messaging

The Remote Procedure Call (RPC) mechanism has been widely used for computation migration. Suppose a process P wants to access file at site A. The access to the file is carried out to site A and could be initiated by an RPC. The process P invokes a predefined procedure on site A. The procedure executes appropriately and then returns the results to P.

Alternatively, process P can send a message to site A. The OS at site A would then create a new process Q whose function is to carry out the designated task when Q completes its execution. It sends the needed result back to P via message system.

Note: In computation migration, a part of the computation of process is normally carried out on different machine.

c. Process Migration: A logical extension of computation migration is process migration. A process may be executed at a computer different from where it is originated. Process migration may be desirable for several reasons:

- **Load Balancing:** The processes may be distributed across the network to even the workload.
- **Computation speedup:** If single process can be divided into a number of sub-processes that may run concurrently on different sites, then the total process turnaround time can be reduced.
- **Software Preference:** The process may require software that is available at only a particular machine.
- **Data Access:** Just as in computation migration, if the data being used in computation are numerous, it may be more efficient to have process run remotely rather than to transfer all data locally.

8. Security: The issues that must be considered in design of security of distributed system are:

- Authentication
- Authorization

Authentication is process of guaranteeing that an entity is what it claims to be.

Authorization is process of deciding what privileges an entity has and making only these privileges available.

By introducing authentication and authorization, the system can be secured.

2.2 THEORETICAL FOUNDATION FOR DISTRIBUTED SYSTEM

Distributed system is a collection of computers connected together via communication network. The processes executing on these computers communicate with one another by exchanging messages over communication channels. The messages are delivered after an arbitrary transmission delay. Thus the two main characteristics of distributed systems are:

1. Absence of global clock
2. Absence of shared memory

2.2.1 Limitations of Distributed Systems

Absence of global clock and absence of shared memory are the limitations that have an effect on the design of distributed system. The system designer must consider all these limitations during design and development of distributed system.

1. Absence of Global Clock

The distributed system environment consists of many computers system with their own system clock. Thus in a distributed system, there exists no system wide clock (global clock).

Time is important and interesting issue because it is used for:

- Time stamping (to know at what time of the day a particular event occurred at particular system).
- Ordering the events.
- Scheduling the processes.

The two approaches for solving the problem of absence of global clock:

- **To have global clock for all the processes in system:** This approach is not useful because two distinct processes can observe the same clock value but at two different instants due to unpredictable message delay.
- **Provide each computer in the system a physical clock and try to synchronize them:** Drift and skew of any physical clock will create problem.

2. Absence of Shared Memory

As all the systems in the distributed system have their own specific physical memory, so the entire system do not have shared memory.

Impact of the Absence of Shared Memory

Due to absence of shared memory global state(Up-to-date state of entire system) is not available to individual process or user. The global state is necessary for:

- Reasoning the system behavior
- Debugging
- Fault tolerance
- Recovery from failure

2.2.2 Different View of the System State

- **Coherent:** A view is said coherent if all the observations of different processes (of different computers) are made at same physical time. This view is said partial view because process is aware of the state of system only for little time.
- **Incoherent view:** This view is related to the local state of all the computers and also any message in transmission state.

Thus the incoherent view encompasses the local views (local state) at all the computers as well the messages in transmission state.

So we can define the incoherent view as global system state.

The coherent view is helpful in distributed system in absence of global clock. But the absence of global clock affects the presence of coherent view of system. Obtaining coherent global state is too difficult in absence of global clock.

Example:

S1, S2: two different sites maintaining account of A and B respectively.

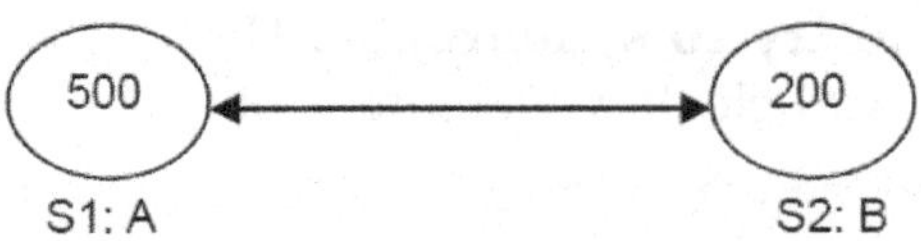

To compute net balance of both accounts, the knowledge of global state of system is necessary.

Let S1 transfer Rs.50 from A to B. During the collection of global state if site S1 records the state of A immediately after debit has occurred and site S2 saves the state of B before the fund transfer message has reached B, the global state of the system will show Rs.50 missing, as shown below:

Clocks, Events, Process States

As distributed system is collection of N processes (N is a finite number) running on different systems, each Process P_i has a state S_i, which in general it transforms as it executes.

As each process P_i executes it takes a series of action, each of which may be either:

- Communication action (message send/receive operation), or
- State transforming action

Note: The state of a process in this discussion includes the value of all the variables within it and value of any object in its local OS environment that it effects, such as file.

An event is defined as the occurrence of a single action that a process carries out as it executes.

Now the history of a process P_i is the series of events that take place within it.

$$\text{history}(P_i) = h_i = \langle e_i^0, e_i^1, e_i^2 \ldots\ldots \rangle$$

2.2.3 Lamport Logical Clock

From point of view of any single process, events are ordered of uniquely by times shown on local clock. However, the clocks on different system may vary, we cannot use the physical time to find out the order of any arbitrary pair of events occurring within it.

Therefore Lamport proposed the logical time that is the relative time between events. To capture the dependencies and to order the events Lamport defined Happened- Before Relation.

Happened Before Relation

We can define happened before relation, denoted → as follows:

- a → b, if a and b are events in same process and a happened before b.
- For any message m

  send (m) → receive (m)

  where send (m) is the event of sending message, and receive (m) is event of receiving it.
- If a, b and c are the events such that

  a → b and b → c

  then a → c

  i.e "→" relation is transitive.

Thus if e and e' are events and if e → e', then we can find a series of events e_1, e_2, e_3, e_4, e_5 ..... e_n occurring at one or more processes either by applying rule 1 or rule 2 of happened before relation. That is, either they occur in succession at same process or there is a message m such that e_i = send (m) and e_{i+1} = receive (m).

Concurrent Events: Two distinct events a and b are said concurrent (denoted by a || b if a $\not\rightarrow$ b and b $\not\rightarrow$ a.

Note: For any two events a and b in a system, either a → b or b → a or a || b.

Example

1. A space-time diagram of two processes P_1 and P_2 is shown below:

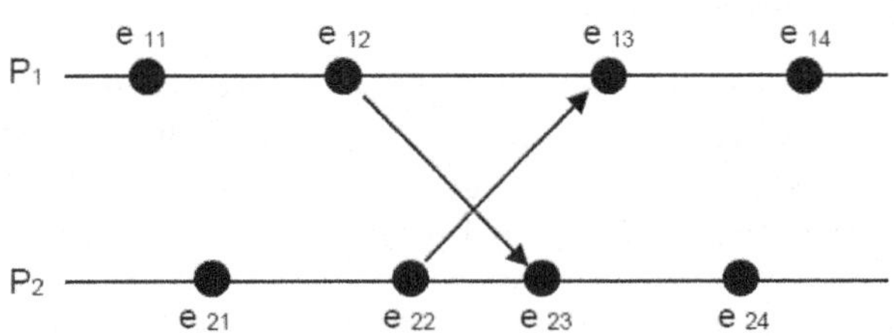

Here, e_{11}, e_{12}, e_{13}, e_{14} are the events in P_1 and e_{21}, e_{22}, e_{23} and e_{24} are events of process P_2. An arrow represents the message transfer between processes. So e_{12} is message sending event from P_1 to P_2 and e_{22} is message sending event from P_2 to P_1 and e_{23} and e_{13} are message receiving event, therefore.

$((e_{11} \rightarrow e_{12}) \parallel (e_{21} \rightarrow e_{22})) \rightarrow ((e_{13} \rightarrow e_{14}) \parallel (e_{23} \rightarrow e_{24}))$

As e_{11} happens before e_{12}, but both are not influencing the occurrence of e_{21} and e_{22} so:

$e_{11} \parallel e_{21}$ and $e_{11} \parallel e_{22}$

$e_{12} \parallel e_{21}$ and $e_{12} \parallel e_{22}$

but after happening of e_{22} and e_{12} both e_{13} and e_{23} may happen concurrently.

2. A space time diagram of two processes.

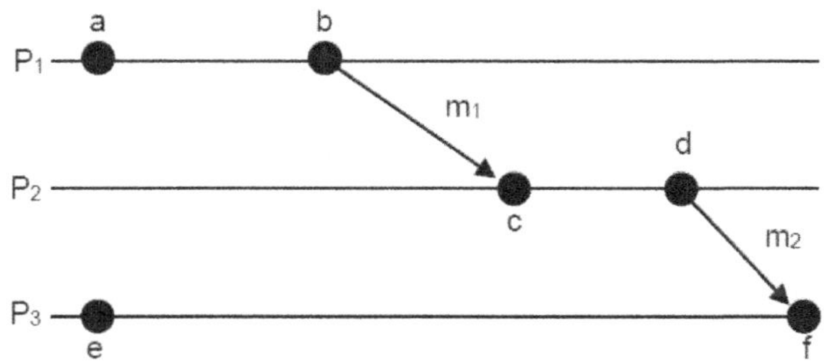

As since $a \rightarrow b$ (since events occur in this order at P_1) and $c \rightarrow d$, $b \rightarrow c$, $d \rightarrow f$

We can say $a \rightarrow f$

But $a \not\rightarrow e$

$e \not\rightarrow a$

so $a \parallel e$

Logical Clock

Lamport invented a simple mechanism by which the happened-before ordering can be captured numerically, called a logical clock. Lamport logical clock is used to timestamp the events (to assign time of occurrence).

A lamport clock is a monotonically increasing software counter, whose value has no relationship with any physical clock. Each process P_i has its own logical clock C_i. The clock C_i can be thought of as function that assigns a number $C_i(a)$ to an event a called timestamp of a) at P_i. As this number has no relation to physical clock or time, this is known as logical clock.

Note: Timestamp of an event is the value of clock when it occurs.

Conditions Satisfied by Logical Clocks

To capture the happened-before relation logical clock follows following conditions:

1. For any two events a and b in process P_i, if a occurs before b then $(a \rightarrow b)$

   $C_i(a) < C_i(b)$
2. If a is event of sending message in P_i and b is event of receiving same message in P_j then

   $C_i(a) < C_j(b)$

Implementation Rules for Logical Clock

1. Clock C_i is incremented between any two successive events in process P_i

   $C_i = C_i + d$ (where $d > 0$)

   If a and b are two successive events and $a \rightarrow b$

   $C_i(b) = C_i(a) + d$
2. When a process P_i sends a message m, then timestamp t_m of this event is associated with m

   $t_m = C_i$
3. If event a is the sending of message m by P_i then on receiving the same message by P_j, C_j is set to value greater than or equal to its present value and greater than t_m.

   $C_j = \max(C_j, t_m + d)$ $(d > 0)$

Note: Message receipt event increments C_j as per rule 1. And that incremented value is being used in above equation.

Following figure shows how the logical locks are updated under Lamport scheme.

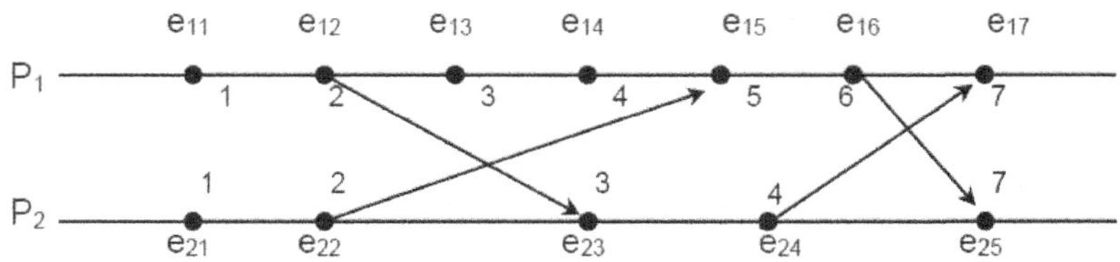

Figure 2.4: Logical Locks Updation

Both the clock value Cp_1 , and Cp_2 are assumed to be zero initially and d is assumed j.

Between two successive events clock value is being incremented by 1. But event e_{25}, which is the receipt of the message (sent on clock value 6). So, Cp_2 (e_{25}) = max (4+1, 6+1) = 7

Similarly,

Cp_1 (e_{24}) = max (4+1, 6+1) = 7

Note: If $a \rightarrow b$

then $C(a) < C(b)$

However, reverse is not always true if the events have occurred in different processes. It is just because those events may be concurrent.

As in the last example:

Cp_2 (e_{21}) < Cp_1 (e_{12})

But $e_{21} \rightarrow e_{12}$ is not true

Totally Ordered Logical Clocks

Lamport happened-before relation $\rightarrow$ defines partial ordering among events. Some pair of the distinct events, generated by different processes, have numerically, identical timestamps. Then, for total ordering of events, identifiers of processes are taken into account to break the tie.

Ordering relation is denote by $\Rightarrow$

i.e., if a is any event at P_i, b is any event P_j then a is ordered before b ($a \Rightarrow b$) only when

$C_i(a) < C_j(b)$ or

$C_i(a) = C_j(b)$ and $i < j$

where i and j are PIP numbers of processes.

Limitations of Lamport Clock

As we know if C(a) < C(b), then a $\rightarrow$ b is not always true.

So if C(a) < C(b), Lamport clock does not guarantee that events a and b are casually related. So just by looking at timestamp of events we cannot say where whether there exists a path between events or not. As in the last figure

$C_i\ (e_{11}) < C_j\ (e_{22})$

But there exists no path between these events so they are casually unrelated events while:

$C_i\ (e_{11}) < C_j\ (e_{23})$

But $e_{11} \rightarrow e_{23}$

As there exists a path between them so they are casually related events.

Therefore, using the timestamp assigned by Lamport clocks, we cannot determine the casual relationship between the two events occurring in different processes, just by looking at timestamp of events.

The reason behind the above limitation is that the Lamport clock system cannot distinguish between the advancement of clock due to local events from those due to exchange of messages between processes.

2.2.4 Vector Clock

Mattern and Fidge developed vector clocks to over come the shortcoming of Lamport clock.

Let n be the number of processes in the distributed system. Each process P_i is equipped with a clock C_i, which is an integer vector of length n. Each process uses its own vector clock to timestamp local events. $C_i(a)$ is referred as timestamp of event a at P_i.

$C_i[\ i\ ]$ i.e., the ith entry of C_i, corresponds to P_i's own logical time.

$C_i\ [\ j\ ]$ where $j \neq i$,, is the P_i's best guess of logical time at P_j.

Implementation Rules of Vector Clock

Initially $C_i[\ j\ ] = 0$ for i, j = 1, 2,...... n.

Clock C_i is incremented between any two successive events in process P_i.

$C_i [i] = C_i [i] + d \quad (d > 0)$

When a process sends a message t_m, it embeds the timestamp t_m of this event with m

$t_m = C_i$

On receiving the same message m, by P_j, processes P_j learns about the more recent clock values of the rest of the processes in the system. So, C_j is updated as:

Example 1:

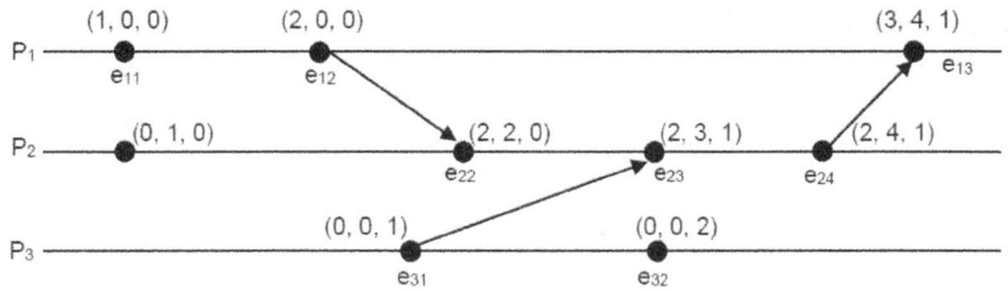

Example 2:

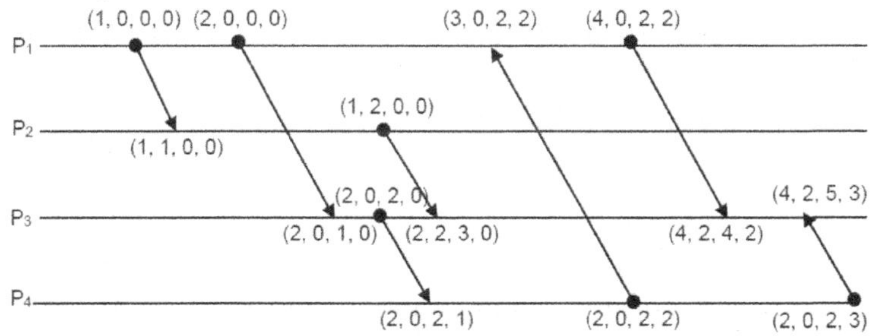

Comparison of Vector Timestamp

For any two vector time stamp t_a and t_b for events a and b, comparison can be made as follows:

Equal

$t_a = t_b$ iff $\forall$ i $t_a[i] = t_b[i]$

for example (1, 0, 3, 2) and (1, 0, 3, 2) are equal timestamps.

Not equal

$t_a \neq t_b$ if $\exists$i $t_a[i] \neq t_b[i]$

for example (3, 3, 2,1) is not equal to (3, 2, 2, 0).

Less than or equal

$t_a \leq t_b$ iff $\forall i\ t_a[i] \leq t_i\ [i]$

for example (3, 2, 0, 1) ≤ (3,1, 0, 1)

(3, 2, 2, 2) ≤ (3, 2, 2, 2)

Less than

$t_a < t_b$ iff $(t_a \leq t_b \wedge t_a \neq t_b)$

for example (3, 2, 0, 1) < (3, 1, 0, 1).

Concurrent

$t_a \parallel t_b$ iff $(t_a \nless t_b \wedge t_b \nless t_a)$

Casually Related Events

Events a and b are casually related if

$t_a < t_b$

or $t_b < t_a$

Otherwise they are concurrent

Thus if $a \rightarrow b$

$t_a < t_b$

Thus the system of vector clock allows us to order events and decide whether they are casually related or not just by looking at timestamps of the event.

2.2.5 Casual Ordering of Messages

Similar to the events, vector clock also facilitate the ordering of messages. The casual ordering of messages deals with the concept of maintaining some casual relationship that holds among "message send" event with corresponding "message receive" event.

In other words, if send $(M_1) \rightarrow$ send (M_2) then every recipient of both messages m_1 & m_2 must receive m, before m_2.

In the distributed system, casual ordering of messages is not guaranteed automatically. The violation of casual ordering of the messages in a distributed system is given in figure 2.5.

The technique for the casual ordering of messages are useful for the applications such as replicated database systems in which it is important that every process in charge of updating a replica receives the updates in the same order to maintain the consistency of database.

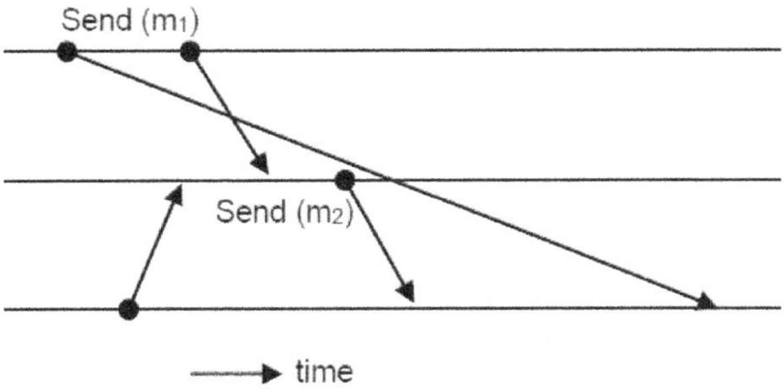

Figure 2.5: Casual Ordering of the Messages

Protocols for Ordering Messages

The basic idea of protocols is to deliver a message to a process only if the message immediately preceding it has been delivered to the process. Otherwise, the message is not delivered immediately but it is buffered until the message immediately preceding it, is delivered.

Birman-Schiper-Stephenson Protocol

1. Before broadcasting a message m, a process P_i increment vector time $VT_{Pi}[i]$ and timestamps m.
   $VT_{Pi} = VT_{Pi}[i] + 1$
   $VT_m = VT_{Pi}$
2. The delivery of same message by P_j is delayed until following both conditions are satisfied:
   - Until it is assured that process P_j has received all the messages from P_i that preceed m.
     $VT_{Pj}[j] = VT_m[i] - 1$
   - Until it is assured that P_j has received all those messages received by P_i before sending m.
     $VT_{Pj}[K] \geq VT_m[K]$ (where $k \neq i$)
   - When a message is delivered at P_j. VT_{Pj} is updated as $VT_{pj}[K] = \max(VT_{Pj}[K], VT_m[k] \; \forall K$.

Example:

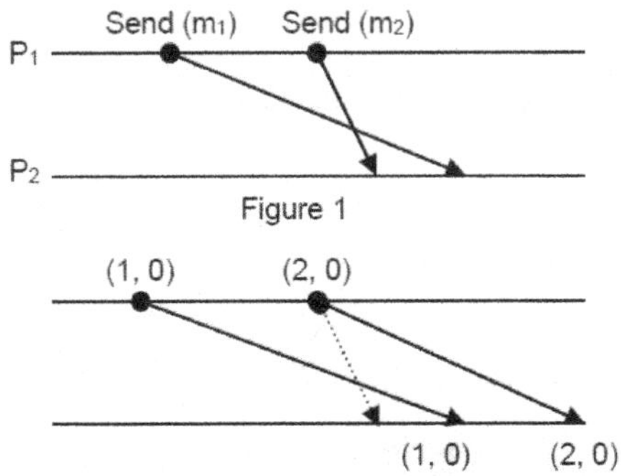

Figure 2.6: Example scenario

The receipt event in **figure 2.6** is delayed $VT_{P2}[1] \neq VT_m[1] - 1$

2.2.6 Global State

"The global state of a distributed computation is the set of local states of all individual processes involved in the computation plus the state of the communication channels."

- process state

  memory state + register state + signal masks + open files + kernel buffers + ...

  Or

  Application specific information like transactions completed, functions executed etc,.

- channel state

  "*Messages in transit*" *i.e.* those messages that have been sent but not yet received

Need for Global States

1. Many problems in Distributed Computing can be cast as executing some action on reaching a particular state
2. e.g.
   - Distributed deadlock detection is finding a cycle in the Wait For Graph.
   - Termination detection
   - Check pointing

- Many more.....

Now the question arises that why global state determination is difficult in Distributed Systems. The very clear reason is:

Distributed State:

Have to collect information that is spread across several machines!!

Only Local knowledge:

A process in the computation does not know the state of other processes.

Consider the following scenario:

S_1 and S_2 are two different sites maintaining account of A and B respectively.

State 1

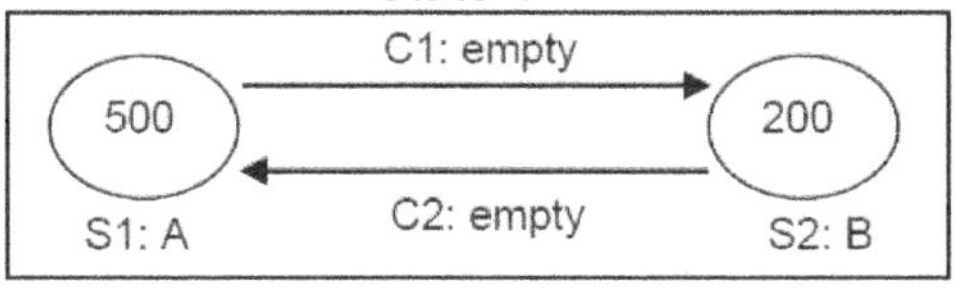

If Rs.50 is transferred from A to B

State 2

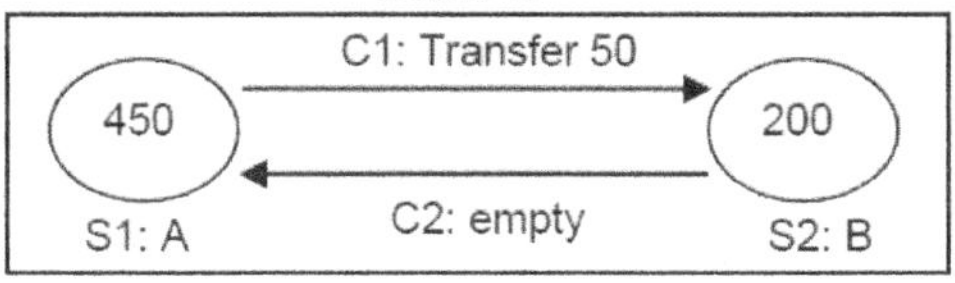

As we know that the global state of a distributed system is the collection of local state of all the sites as well as local state of all the sites as well as local states of communication channels.

So if local state of S_1 is recorded when ever system was in state 1 and local states of C_1 and C_2 and S_2 are recorded if the system was in state 2, then global state will be

A has balance 500

B has balance 200

C_1: transfer 50

C_2: empty

So extra Rs.50 is in global state Therefore this global state is inconsistent.

Reason for this Inconsistency

A's local state was recorded before the message was sent and channel C_1's local state was recorded after the message was sent.

Therefore, The global state may be inconsistent if

(If site A is sending message to B.)

No of messages sent by A along channel before A's local state was recorded $<$ No of messages sent by A along same channel before channel's state is recorded

$n > n'$ ...(1)

If channels states are recorded when the system was in state 1 and A's and B's local state is recorded when ever system was in state 2, then global state will be:

A has balance 450

B has balance 200

C_1: empty

C_2: empty

Therefore Rs.50 is deficit in global state (inconsistent global state)

The reason for this inconsistency

$n > n'$

Hence for consistent global state $n \nless n'$ and $n \ngtr n'$

$n = n'$ ...(2)

Till now our consideration was at sender end, now if we think at receiver end, we can determine consistent global state requires:

No of messages received along channel before B's state is recorded $<$ No of messages received along the channel by B before channel's state was recorded

$m = m'$ ...(3)

Since in number system, the number of messages sent along the channel can be less than the number of messages received along that channel so:

$n' \geq m$ ...(4)

Thus from equations (4) and (2)

For consistent global state

$n \geq m$

Example:

Consider the events occurring at process P_1 and P_2

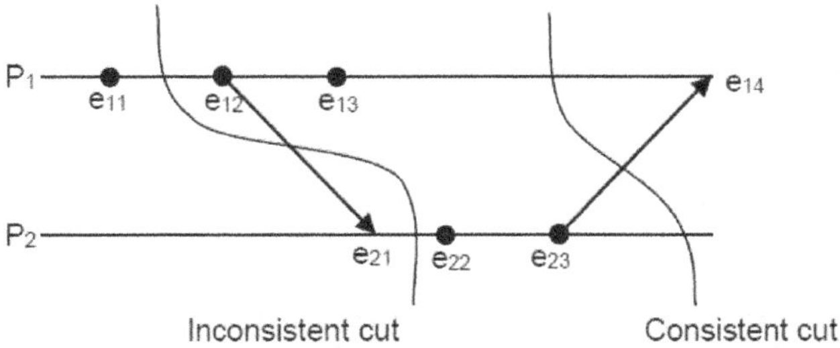

- The left most cut is inconsistent because $n \ngeq m$, $0 \ngeq 1$.
- The right most cut is consistent because $n \geq m$, $0 \geq 0$

Mathematical Representation of Global State

A global state GS of a system is a collection of local states of its sites.

$GS = \{LS_1, LS_2 \ldots\ldots LS_n\}$

Where LS_i = local state of site S_i at any time

n = number of sites in system.

Let,

send (m_{ij}) – send event of message m_{ij} by S_i to S_j.

recv (m_{ij}) – receive event of m_{ij} by S_j

time (x) – time at which state x was recorded

time (send (m)) – time at which send (m) occurred.

So for a message m_{ij} sent by S_i to S_j

- send (m_j) $\in LS_i$ iff time (send (m_{ij})) < time (LS_i)
- recv (m_j) $\in LS_j$ iff time (recv (m_{ij})) < time (LS_j)

Now, in a global state a message will be said in transit when its sending event has been included in local state of sender and its receiving event is not included in the local state of receiver.

transit $(LS_i, LS_j) = \{m_{ij} | \text{ send } (m_{ij}) \in LS_i \wedge \text{ recv } (m_{ij}) \notin LS_j\}$

If there is any message m_{ij} from S_i to S_j, where sending event is not included in local state of sender (LS_i) and receiving event is included in local states of receiver (LS_j) then GS combining these local states ($LS_i, LS_j \in GS$) is inconsistent.

Inconsistent $(LS_i, LS_j) = \{m_{ij} | \text{ send } (m_{ij}) \notin LS_i \wedge \text{ recv } (m_{ij}) \in LS_j\}$

Consistent Global State

A global state $GS = \{LS_1, LS_2 \ldots\ldots LS_n\}$ is consistent iff

Inconsistent $(LS_i, LS_j) = \phi$

$\forall_i, \forall_j$

Thus in consistent global state, for every received message a corresponding send event is recorded in global state.

In consistent global state, there is at least one message whose receive event is recorded but its send event is not recorded in global state.

Note:

A global state is said transitless iff

transit $(LS_i, LS_j) = \phi\ \forall_i, \forall_j$.

i.e., for every sent message, corresponding receive event is recorded in global state. So all communication channels are empty.

- $\{LS_{12}, LS_{23}, LS_{33}\}$ is consistent global state.
- $\{LS_{11}, LS_{22}, LS_{32}\}$ is in consistent global state.

It is also transitless global state.

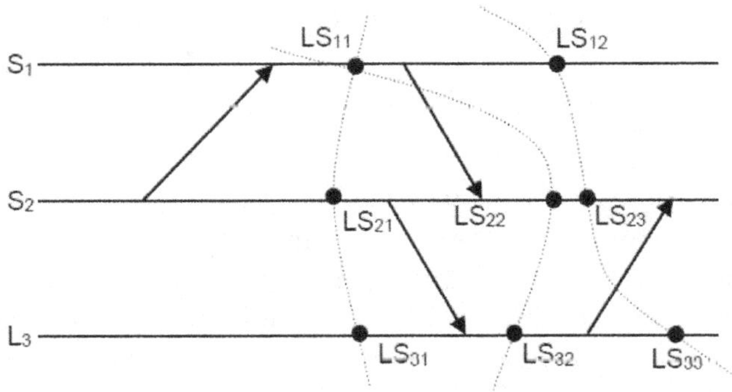

Note: A global state is strongly consistent if it is consistent & transitless. {LS_{11}, LS_{21}, LS_{31}} is strongly consistent.

Example 2:

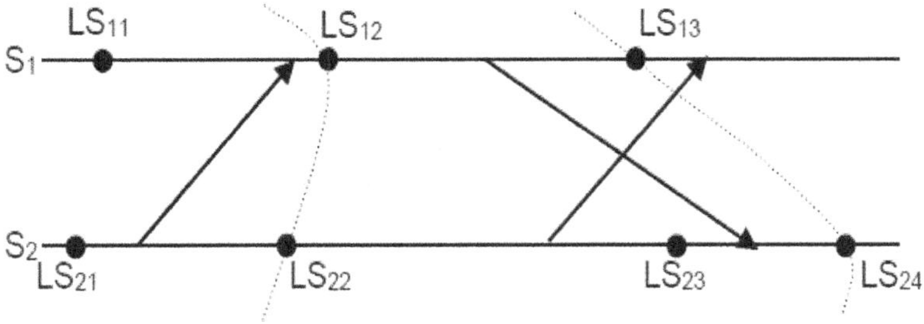

- GS = {LS_{12}, LS_{22}} is consistent and transitless so strongly consistent.
- GS = {LS_{13}, LS_{24}} is consistent.
- GS = {LS_{12}, LS_{24}} is inconsistent.

2.2.7 Chandy – Lamport's Global State Recording Algorithm

- **Marker sending rule for process P:** P sends a marker message along all its outgoing channels C after it records its state and before it sends any other messages.
- **On receipt of a marker message by Q along a channel C:** If Q has not recorded its state

  begin

  Record the state of C as empty

  follow the marker sending rule

  end

Else

Record the state of C as sequence of messages received along C since it has recorded its state excluding marker.

Algorithm in Action

The global state recording algorithm can be initiated by any process by executing marker sending rule.

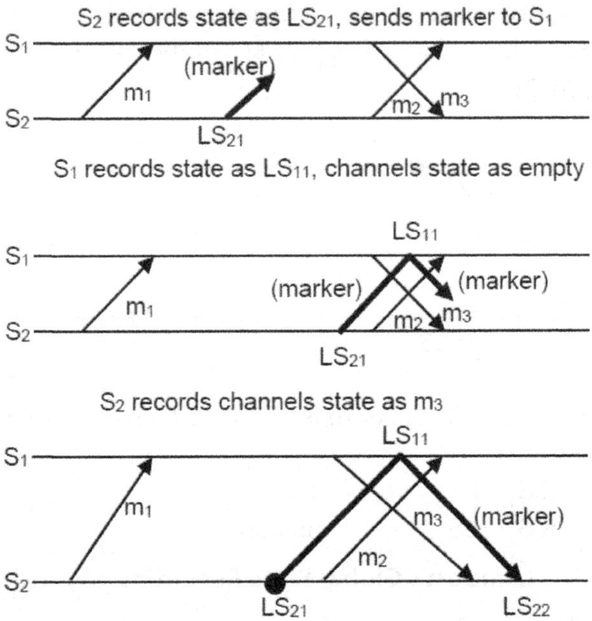

2.2.8 Termination Detection

Generally we adopt the usual model of distributed system consisting of n processes P_1 ...... P_n running on various sites and communicating via synchronous messages.

A distributed computation consist of set of co-operating processes which communicate with each other by exchanging messages. In the case of distributed computation, it is important to know when the computation has terminated.

Some facts related to distributed computation consisting of P_1 ...... P_n.

- A process is always in one of two states:
  - Active
  - Passive
- A Passive process is not engaged in any activity of its own but is prepared to respond with a value requested by other.
- An active process may activate other passive processes by sending a message to it. (Computation message)

- At any time, an active process may change its state to passive (to become active again, it must be reactivated by an active process)
- A computation is said to be terminated if and only if all the processes are idle and there are no message in transit.

Huang's Termination Detection Algorithm

One of the co-operative process monitors the computation and is called controlling agent. Initially all processes are idle, the controlling agent's weight is set 1 and weight of rest processes is set 0.

The computation starts when controlling agent sends a computation message to one of the process. Any time a process sends a message the process's weight is split between itself and the process receiving the message. The weight received along with a message is added to the weight of process. Thus, the algorithm assigns a weight W to each active process and to each message in transit.

The weight of processes lies between 0 and 1.

$0 < w \geq 1.$

The weights are assigned to processes in such a manner so that sum of all weights will be equal to one

$\sum w = 1.$

When the computation is completed, the process sends its weight to controlling agent. Controlling agent adds the received weight to its weight. The weight of the controlling agent is main measurement of termination detection.

If the weight of controlling agent is one after completion of a computation, the computation is assumed terminated.

The following notations are used in this algorithm:

B(w): Computation message sent by controlling agent and having weight w.

C(w): Control message sent by process to controlling agent and having weight w.

The Haung's Termination detection algorithm uses above notions for applying following steps:

Step 1: The controlling agent or an active process having weight w may send computational message. During this following calculation takes place:

Derive w_1 and w_2 such that

$w = w_1 + w_2$ where $w_1 > 0$, $w_2 > 0$

$w = w_1$

Send B (w_2) to P.

Step 2: When process P having weight w receives computation message B (w_2), It becomes active if it is idle and performs following:

$w = w + w_2$

Step 3: An active process having weight w becomes idle at any time by following action:

Send C (w) to controlling agent

w = 0;

Step 4: On receiving C (w), the controlling agent having weight w, takes following action:

w = w + w

If w = 1, Computation is terminated.

2.3 DISTRIBUTED MUTUAL EXCLUSION

2.3.1 Introduction

When two or more processes compete for the use of system resources, there is a need for a mechanism to enforce mutual exclusion. Suppose that two or more processes require access to a single non-sharable resource, such as printer. We will refer to such a resource as critical resource and portion of program that uses it as critical section of the program.

The successful concurrency among processes requires ability to define critical section and to enforce mutual exclusion (i.e., only one process should be allowed to be in its critical section at a time).

2.3.2 The Classification of Mutual Exclusion Algorithms

In single processor systems, critical section regions are protected using semaphores, monitors and similar contracts. But in distributed system, algorithms for enforcing mutual exclusion may be classified as:

- Centralized Algorithm
- Non-token based algorithms.
- Token-based algorithms.

A. Centralized Algorithm

One node is designated as control node and controls access to all shared objects. The control node is also known as coordinator.

Whenever a process wants to enter critical section, it sends a message to the coordinator telling which critical region it wants to enter and asking for permission. If no other process is currently in critical region, the coordinator sends back a reply granting permission (figure a below). When reply arrives the requesting process enters the critical section. Now if at this time any other process wants to enter same critical section (i.e., to access same shared resource), and asks for permission to enter the same critical region, as the coordinator knows that a different process is already in critical region, so it cannot grant permission. So it queues that request and refrains from replying thus blocking that process. (figure b and c below).

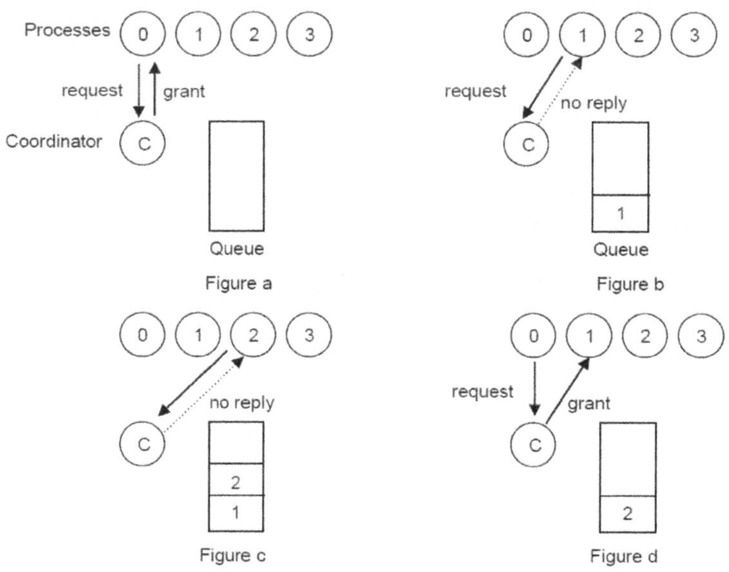

Figure 2.7: Centralized Algorithm working

When the first process exits the critical section, it sends the message to coordinator releasing its exclusive access (figure d below). The coordinator takes first item from queue of deferred requests and sends that process a grant message.

Note: Each process sends the request to coordinator with its process-id and node-id

Drawbacks:

- The coordinator is single point of failure so if it crashes, the entire system may go down.
- In a large system, a single coordinate can become performance bottleneck.

B. NON-TOKEN BASED ALGORITHM

Lamport Algorithm

Every site keeps a queue, named as $request\_queue_i$, which contains mutual exclusion requests ordered by their time stamps.

Requesting the Critical Section

- When a site S_i wants to enter the CS, it sends a REQUEST (t_s, i) message to all the sites and places the requests in its $request\_queue_i$
- When a site S_j receives REQUEST (ts_i, i) message from S_i, it returns a timestamped reply message to S_i and places the S_i's request on $request\_queue_j$

Executing the Critical Section

The Site S_i enters the critical section when following conditions hold:

- S_i has received a message with timestamped larger than (ts_i, i) from all other sites.
- S_i's request is at top of $request\_queue_j$

Releasing the Critical Section

- Upon existing the CS, site S_i removes its requests from top of its request queue and sends a timestamped RELEASE message to all the sites.

- When a site S_j receives a RELEASE message from site S_i, it removes S_i's request from its request queue.

Example: S_1 and S_2 are making requests for CS and send request to all other sites suppose clock values at time are 2 and 1 for S_1 and S_2 respectively.

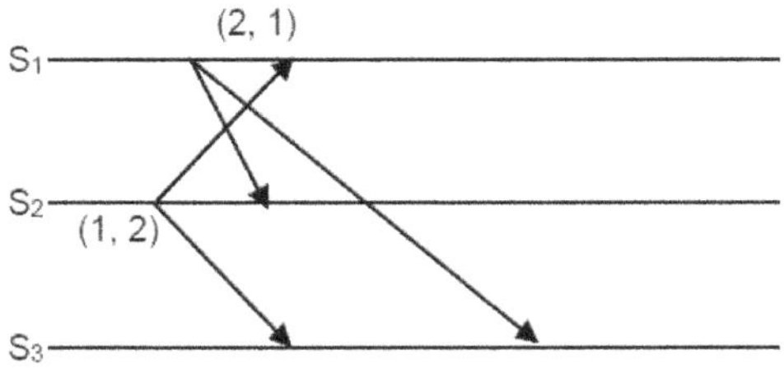

On receiving that requests, sites put the requests in their request_queue and send timestamped reply

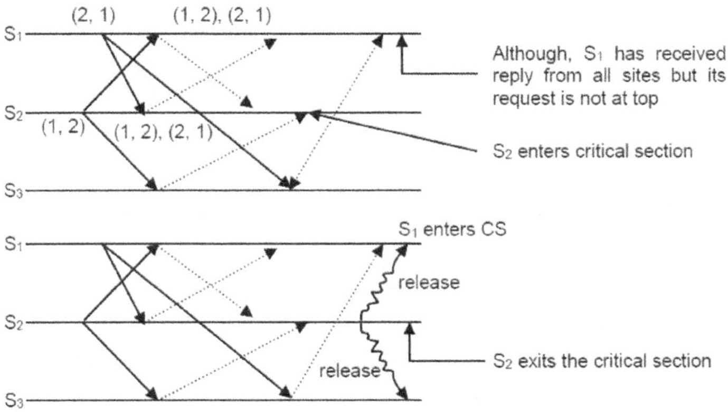

Drawback: This algorithm requires 3 (N – 1) messages per Critical Section invocation where

N = No of sites

(N – 1) Request

(N – 1) Reply

(N – 1) Release

The Ricart Agrawala Algorithm

This algorithm is the optimized form of Lamport Algorithm. It is based on release message send by the site as reply. The reply message is received from many sites as distributed system. This algorithm has the following steps to perform:

- **Step 1:** In a distributed system site S1 needs to enter the critical section. Then it sends the request message with the stamp to all other sites in the system.
- **Step 2:** The site S2 receives the request message from S1 if the time stamp of S1 is smaller than S2 otherwise the request is discarded.
- **Step 3:** When the site S1 enters the critical section. It starts executing but before this entry the site S1 has received all the reply messages from all other sites.
- **Step 4:** The site S1 exit the critical section & then sends the reply message to all other sites.

The reply messages are stored in the form of a block. The site with smaller time stamp has the ability to send the reply message. This site is called higher priority site for a critical state. The reply message is generated necessarily on the exit of the site from the critical state. This problem is continuous again and again.

The following theorem proves the correctness of the Ricart Agrawala Algorithm.

Theorem: According to Recard Agrawala Algorithm, there is achieved mutual exclusion.

Proof: Suppose the two sites S1 & S2 are executing the critical state concurrently. The request of site S1 has a similar time stamp, so the request of site S1will execute first. Request of S2 is received after the request of site S1. Both the sites can execute the critical section concurrently only when the S1 sends the reply message to site S2 & before the exit of S1 from critical section. Almost all the time, S2 will execute its critical section after the site S1, because the request of S2 has low priority. In this way the Recard Agrawala Algorithm achieves the mutual exclusion.

According to the Recard Agrawala Algorithm, there may be many pair of sites requesting for the execution of the critical section. In this pair one site have higher priority & other have lower priority request, Then always the higher priority request defeats the lower priority request. The request with higher priority will only be successful in receiving all the acknowledgement & reply messages.

Performance Report of Recart Agrawala Algorithm

The message per critical state are required 2(N-1) in number for the execution of critical state. These numbers of messages are in the form of request messages & reply messages. The number of the request messages is equal to the number of reply messages are received & in turn, (N-1) reply messages are produced.

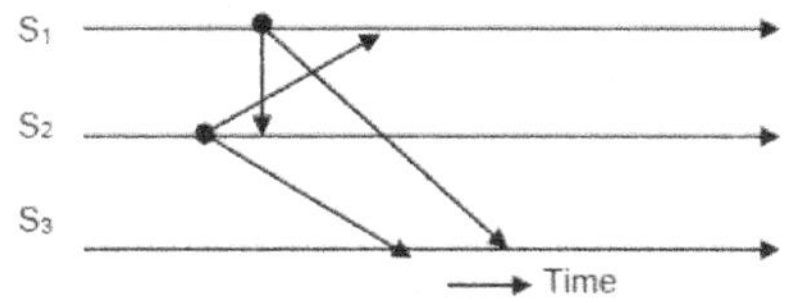

(Request for Critical state by Site S_1 and S_2)

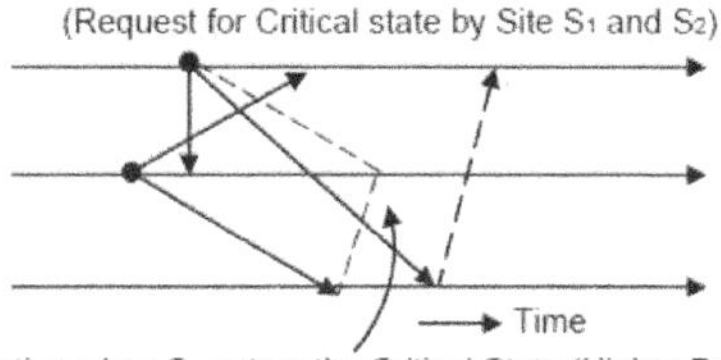

Situation when S_2 enters the Critical State (Higher Priority)

(Arising of Critical State for site S_2

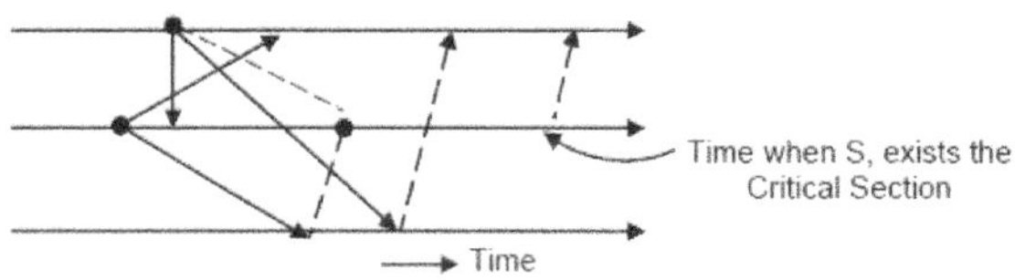

(Exit from CS and Reply messages by S_2

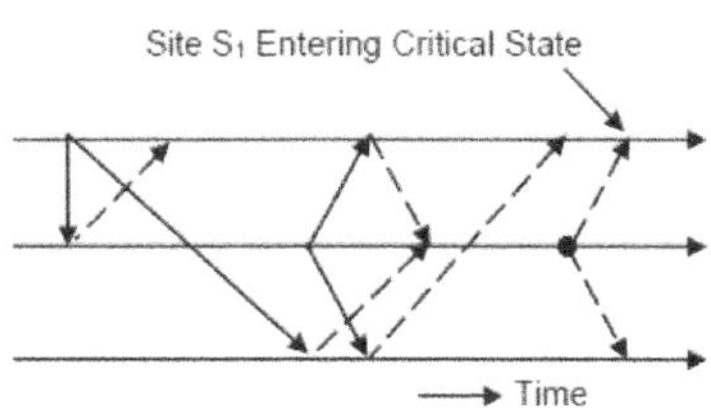

Site S_1 Entering Critical State

An improvement to the Recard Agrawala Algorithm was proposed by ***Roucairol*** and ***Carvalho***. This improvement observed that the site S1 receives the reply message from the site S2. The authorization remains valid till the site S1 sends a reply message to site S2. Thus the site requests the permission for various sites & requires 0 to 2(N-1) messages for one critical state execution.

Mackawa's Algorithm

Mackawa's Algorithm differs from all other algorithms very much only in its approach used. According to the Mackawa's Algorithm, the site in the distributed system does not make a request from any other site.It creates the subset of the site required in the form of request set. Between the pair of sites, the mediator site is present which conflicts between a pair of sites (S1, S2) for the mutual exclusion.

According to Mackawa's Algorithm, a site in the distributed system can send only one reply message at a time after receiving the release message. When a reply message is send by a site after a little time, it receives a release message to ensure the delivery of reply message successfully. Then and only then the site becomes able to send the next reply message. It means, the all other sites are locked immediately when a site (S1) tries to execute the critical state. The other sites are locked in the exclusive mode i.e. no other site can enter the critical section at that time.

In Mackawa's Algorithm the request set of each site is chosen to satisfy the following condition.

The request is made when

C1 : $\forall i \; \forall j : 1 \leq \; i, j \leq n :: R_i \cap R_j \neq 0$

The other situation for the request sets are

C2 : $(\forall \; n : 1 \leq N :: S_n \in R_n)$

C3 : $\forall \; n : 1 \leq n \leq N :: \; | \; R_n | = A$

C4 : $1 \leq n, \; m \leq N$

There may be at least one common site between requests of two sites. Every pair of sites has common site that produces a conflict between the sites of a pair. A site in the pair have only one reply message at a time. A site in the pair have only one reply message at a time. A request is permitted to execute when it is not permitted by any other site.

These conditions can arise for the request of any site (S1 and S2). The algorithm requires the delivery of the message in their sending order between the pair of site.

Almost in all situations, condition C1 & C2 arise and are useful for the application of Mackawa's Algorithm. The condition

C3 states that all sites must have an equal work to establish the mutual exclusion. The condition C4 states that all the sites in distributed system have equal responsibility granting the other permissions to other sites.

The following steps are used by the Mackawa's Algorithm:

- **Step 1:** The site S1 requests to access the critical state and produces the request message REQUEST(1) to all the sites present in request set R.
- **Step 2:** The site S2 receives the request message REQUEST(1) . Then the site S2 sends the reply message REPLY(2) to the site S1. Some times the REQUEST(1) is stores in the queue for further use.
- **Step 3:** The site S1 receives the reply messages from all other sites present in the request set R. Then the site S1 enters the critical section for the execution of critical state.
- **Step 4:** When the execution of critical state is over for site S1 then the site S1 sends the release Message RELEASE(1) to all other sites of the request set.
- **Step 5:** The site S2 receives the release message RELEASE(1) from the site S1. Then S2 sends a reply message to the next waiting site in the request queue till the request queue becomes empty. If the queue is empty, the state of site is updated.

C. Token Based Algorithm

In token-based algorithms, a unique token is shared among all sites. A site is allowed to enter its CS if it possesses the token.

Basic Idea: A token is a special type of message that is passed around the system. Possession of the token entitles the holder to enter the CS. Since there is only a single token, only one process can be in its CS.

Depending on the way a site carries out its search for the token, there are numerous token-based algorithm.

Ring-Based Algorithm

We assume that the processes in the system are logically organized in ring structure. The physical communication network does not need to be a ring as long as processes are connected to each other; it is possible to implement a logical ring.

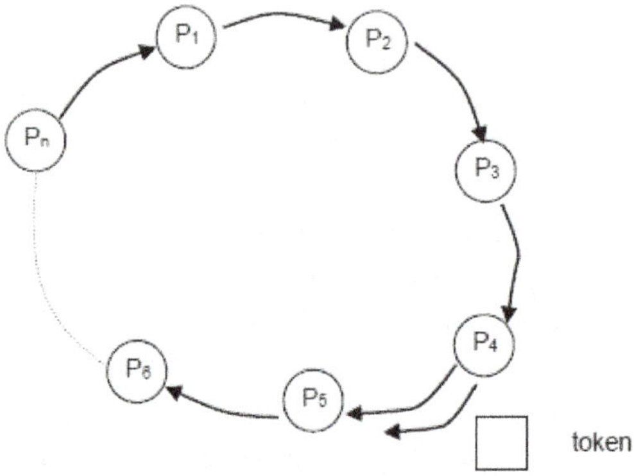

Figure 2.8: Idea of Ring based algorithm

To implement the mutual exclusion, we pass the token around ring. When a process receives the token it may enter its CS, keeping the token. After process exists CS, the token is passed around again. If the process receiving the token does not want to enter CS, it just passes the token to its neighbor.

Drawbacks

- If a token is lost, it must be regenerated but detecting that it is lost is difficult, since the amount of time between successive appearances of token on network is unbounded. The fact that token has not been spotted for an hour does not mean it has been lost (somebody may be still using it).
- If any process fails, the ring is broken.

Singhal's Heuristic Algorithm

According to this algorithm each and every site in the distributed system contains the information about the other site-states present in the system. This information keeping system is helpful in finding the set of sites having token with themselves. Heuristic means 'based on experimental facts'. The algorithm is called Heuristic Algorithm because the set of sites is selected heuristically.

The requests for token are sent to the set or subset of sites. The requesting site sends a request message to another site that is holding the token. Sometimes the site to which request is made does not hold the token but it will be going to catch the token further. So the request is made only to the site, which is having the token or will be having the token in future. Hence the set or subset of sites ensures the presence of token currently or in future. This fact is necessary to consider during the design of algorithm.

In this situation the array is useful to be the better data structure. The arrays are to hold the information about the sites in the system. These arrays store the states of each and every site in the system.

The information about the sequence number of each site and its highest possible sequence number is also kept within the array.

In the similar way the token s maintain the array structures for holding the token state and token sequence numbers.

The sequence numbers of the site and token are useful to get the details about the pending requests and outdated requests.

We know that a site may be in different states during process in the distributed system; such as requesting states, executing states, idle state, etc. The Heuristic algorithm selects every requesting site that is trying to request for the critical state. If the two sites are requesting the same token concurrently one of them will always send a token request message to the other site. The Heuristic algorithm works in the following manner:

Step 1:

- If the requesting site (S1) for critical state does not have a token then its state is initialized to **Requesting State.**
- Its sequence number increases by one.
- Then the request message is sent to all other sites of the systems.

Step 2:

- When a site (S2) receives the request message the sequence number of the site is checked.
- If the sequence number of the receiver site is greater than that of requester site the message is discarded due to outdate.
- Otherwise the sequence number of S1 is taken by S2 and then after processing the token by S2 it is send to the site S1.

Step 3:

- After receiving the token S1 executes the critical section with its state set as executing.

Step 4:

- When the execution of critical state is over by site S1 the state and token both are set to non-initialized state for further use.

State 5:

If (∀x::S[x]=Non-initialized),
Then set S[x]:=idle token
Else send the token to site S2.

This algorithm works very accurately with the selection set for receiving the token. After the receiving of token the site finishes a critical section. The token should not be granted to a site waiting for the token then the token can be assigned to the same site again. If there is a site (A) requesting for the token that has already accessed the token then it is checked for other sites to request for token and if any (B) the token is given to the site (B) first rather than (A).

The performance of the Heuristic algorithm depends upon the accessing of token by a site. A site accesses the critical section without communicating with all other sites of the system. So the traffic for the request message is low because each site sends he request message to only those having token or will be having token in future. This checking can be easily done by finding the sequence number of the token and site. When a request is made the sequence number increases by one every time. A little synchronization delay may also be found. This algorithm is heuristic in nature. So the CS requests changes according to the environment.

Suzuki Kasami Broadcast Algorithm

According to this algorithm the requesting site broadcasts a **Request Message** for the token to all other sites, if he site is not having a token to enter the critical section. In the system all sites receives the broadcast message and site having the token it to the requesting site. It might be possible that the site in **Executing State** receives the request message then it sends the token to the requesting site only when it exists from the critical section. The site having a token can enter its critical state again and again until a request from any other site is received.

These algorithms differentiate between the current request and outdated request easily. It determines the site requesting number.

The outdated request (old request) it determines with the help of a sequence number that indicates the site's requesting number.

The request sequences are managed by an array and have a maximum limit for the sequence number. When the request message is found then the sequence number is incremented by one. If the sequence number of the received message reaches higher than the maximum sequence number then the request is made outdated. It can be shown as follows:

RN[x]:=Max(RN(x),n)	State up to which request is received
If RN[x] > n	Then request received.

Here RN is the array holding the sequence number for the requests and n is the maximum sequence number possible.

The algorithm is processed by the following steps:

- **Step 1:** If a requesting site (S1) does not have a token the sequence number is incremented by one then the request message with the sequence number is send to all other sites.
- **Step 2:** When a site (S2) in the system receives the message the (S2) sends the token to (S1) only if the sequence number (S2) one less than the sequence number of (S1).

- **Step 3:** The site S1 then executes the critical section after receiving the token from site S2.
- **Step 4:** After execution of critical section by S1 the sequence number of token increases by one.
- **Step 5:** If there is no site in the request queue the token is send to the top site ID in the queue.

After the execution of the critical section a new site is given the token, which has the current request. So there might be situation of starvation developed because the old sequencing sites are made outdated. Hence this situation is very critical to handle by Suzuki Kasami Broadcast Algorithm.

The following theorem is used to check the correctness of the theorem.

Theorem: A requesting site enters the critical section in finite times.

Proof: The request message of a site (S1) reaches to other sites in finite times. Any site will have the token in a finite time. The request of S1 will be placed in the token queue. There may be maximum (N-1) requests in the token queue. The Suzuki Kasami Broadcast Algorithm is very simple and efficient to achieve the mutual exclusion. This algorithm requires 0 or N messages for critical section.

Raymond Tree Based Algorithm

In this algorithm, sites are logically arranged as directed tree such that edges of the tree are assigned directions towards the site that has token. Every site has local variable holder that points to an immediate neighbor on the directed path to root node. At root site holder points to itself.

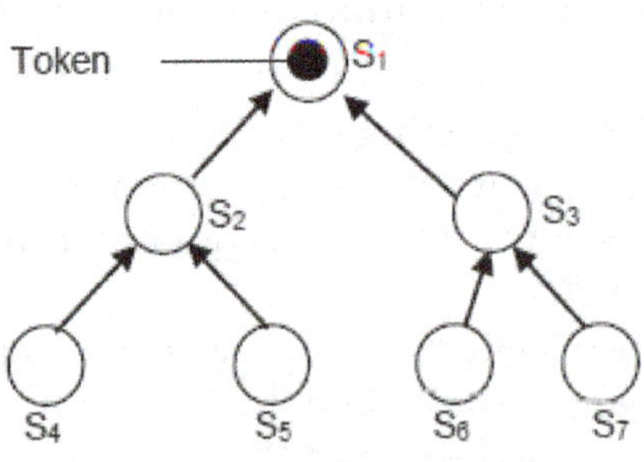

Every site has FIFO queue which stores the request of those neighboring sites that have sent a request to that site but not yet been sent token.

Algorithm

Requesting Critical Section

- When site wants to enter Critical Section, it sends a REQUEST message to the node along the directed path to the root, provided it does not hold the token and request_q is empty. It adds its request to its request_q.
- When a site on the path receives this message, it places the REQUEST in its request_q and sends a REQUEST message along the directed path to the root.
- When a root site receives the REQUEST message, it sends the token to the site from which it received the REQUEST message and sets its holder variable to point at that site.
- When a site receives the token, it deletes the top entry from its request_q sends the token to the site indicated in this entry, and sets its holder variable to point at that site.

Executing the Critical Section

A site enters the Critical Section when it receives the token and its own entry is at top of its request_q. In this case, the site deletes the top entry from its request_q and enters Critical Section.

Releasing the Critical Section

If the request_q is nonempty, then it deletes the top entry from its request_q , sends the token to that site and sets the holder variable to point at that site.

EXERCISE

Q1. Discuss in short about the limitation of distributed system.

Q2. What problem comes in distributed system due to absence of Global Clock?

Q3. How the absence of shared memory affects the working of distributed system?

Q4. Write about the working of Lamport Logical Clock.

Q5. Discuss briefly about the casual events in distributed system.

Q6. Discuss one algorithm for Global State recording.

Q7. What are the Vector Clocks? What are the advantages of Vector Clocks over Lamport Clock? For the space time diagram shown below, obtain the vector time stamp of various events.

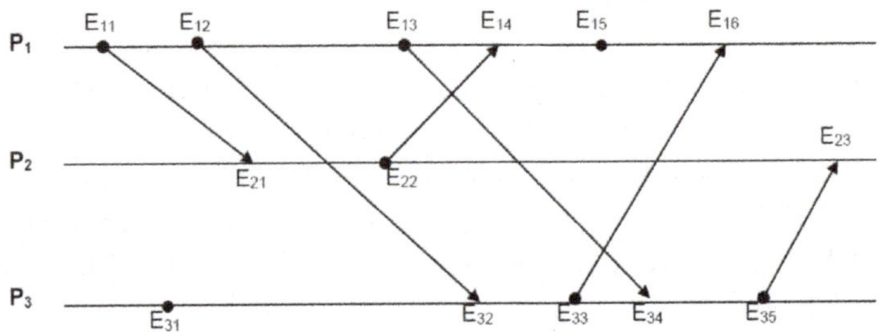

Q8. What is the need of termination detection in distributed systems? Write the Haung's algorithm for process termination detection

Q9. What do you understand by global state of an application? Discuss the Chandy-Lamport's Algorithm for global state recording.

Q10. Define the global state of distributed system. What are the differences in consistent global state and transit-less global state? Consider the following diagram depicting local states for various sites:

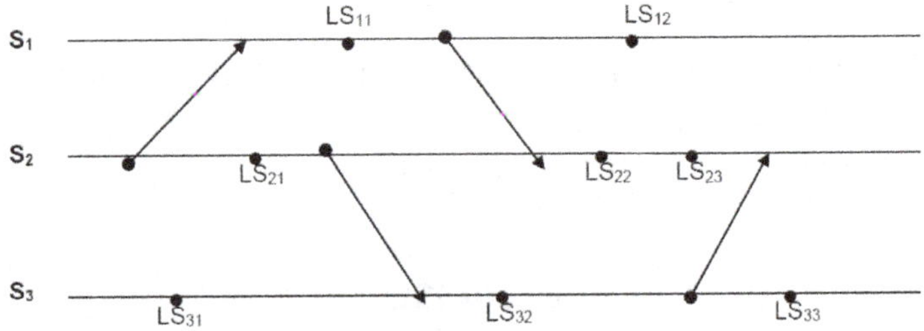

Q11. Find any two consistent global states, transit-less global sate and a strong consistent global state.

Q12. What do you mean by casual ordering of Messages? Discuss the salient features of broadcast based protocol that makes the uses of vector clock which ensures casual ordering of messages.

Q13. Define the mutual exclusion. Differentiate between the approaches used in the token-based and non token-based algorithms for mutual exclusion.

Q14. Differentiate Mutual Exclusion in context with single-computer systems and distributed systems. Discuss how Maekawa's Algorithm guarantees Mutual Exclusion.

Q15. Show that in the Recart-Agrawala algorithm, critical section is accessed according to the increasing order of timestamps. Does the same hold true in Lamport algorithm.

Q16. Ring Algorithm is a Token-Based Algorithm. Discuss its limitations. Discuss also how they are overcome by Raymond's Tree based algorithm?

3

Distributed Deadlock Detection and Agreement Protocol

3.1 DISTRIBUTED DEADLOCK DETECTION

3.1.1 Introduction

The deadlock is defined as the permanent blocking of the processes. That is, a set of processes is waiting for an event that is never going to occur.

Types of deadlock in distributed system:

- **Resource Deadlock:** The resource deadlock occurs when a set of processes is competing for the resources such as data object, I/O resources, etc., and no one is going to proceed.
- **Communication Deadlock:** In the communication deadlock, messages are the resources, for which processes wait. Deadlock occurs, if each process in a set is waiting for message from another process in the set and no process in the set ever sends the message.

3.1.2 Deadlock Handling Techniques in Distributed System

Similar to the single processor system basic strategies for deadlock handling are:

- **Deadlock Prevention:** Statically make deadlocks structurally impossible.

- **Deadlock Avoidance:** Avoid deadlocks by allocating resources carefully.
- **Deadlock Detection:** Allow deadlocks to occur, detect them and try to recover.

1. Deadlock Prevention: We know that if the following four conditions hold at same time in a system the deadlock occurs.

- Mutual exclusion
- Hold & wait
- No Pre-emption
- Circular wait

The basic philosophy of deadlock to prevention is to deny at least one of four necessary conditions for deadlock.

In the distributed environment deadlock can be prevented as follows:

- The circular wait condition can be prevented by global ordering of resources. That is, all resources in entire distributed system are assigned a unique number and a process may request a resource (at any site) with unique number i only if it is not holding a resource with unique number greater than i.
- The hold-wait condition can be prevented by requiring that a process request all of its required resources at one time, and blocking the process until all the requests can be granted simultaneously.

This approach may also lead to deadlock.

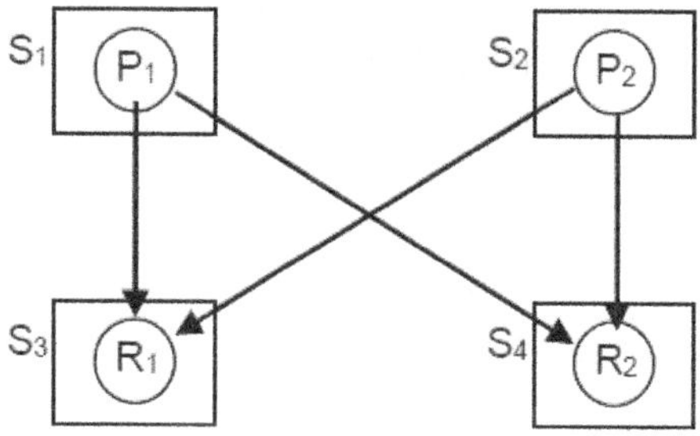

If P_1 and P_2 simultaneously request resources R_1 and R_2 located on S_3 and S_4 respectively and S_3 grants all requested resources by P_1 (i.e., R_1) to it and S_4 grants R_2 to P_2. There will be deadlock.

2. Deadlock avoidance: In the Distributed System, many systems work simultaneously. All the process and resources included in the distributed system make the Global State of the distributed System. The Concept of Global State is useful for a distributed system having all the processes and resources together. The Deadlock avoidance approach focuses on avoiding the deadlock than its removal.

According to this approach, a resource is granted to a process, iff the entire Global State of the system remains safe. The deadlock avoidance is applied o entire system and do no for a single process. So the knowledge of Global State is necessary for his approach. But it is not easy task for the every site to be aware of the Global State of the system. In the distributed system, the processing us distributed across various sites. So the global state of a process or entire system is not known to a particular system. The global state includes the large information about all the processes and resources of the system.

The ***Safe Global State*** is needed for deadlock avoidance. The checking of the safe global state is done efficiently by the system. All the sites are not to be allowed to check for the safe global state concurrently because it will lead to *Net Unsafe Global State.*

3. Deadlock Detection: The deadlock detection is the process of determining that deadlock exists and identifying the process and resources involved in deadlock.

In this approach, resources are granted to requesting processes without any check. Periodically system state is examined to determine if a set of processes is deadlocked.

The deadlock detection involves two main issues:

- Maintenance of Wait for Graph, which describes the resource allocation state.
- Search for cycle when each resource has only onc unit or for knot when resources have more than one unit.

Depending upon the manner in which Wait for Graph is required is organized in distributed system and search for cycle/ knot is carried out, the detection approaches can be classified as:

- Centralized control

- Distributed control
- Hierarchical control

a. Centralized Control: With centralized deadlock detection approach, one site, designated, as control site is responsible for deadlock detection. The control site has responsibility for constructing global wait for graph and searching for the cycle/ knot. The global Wait for Graph is union of all local Wait for Graph.

The recovery form deadlock is simple in this algorithm because control site has complete information of system and it also has information of deadlocked processes. Thus, it can resolve the deadlock.

Note: The nodes of the local wait-for graph correspond to all the processes (local as well as non local) currently holding or requesting any of the resources local to the site.

When P_i in site S_1, needs a resource (on S_2) held by P_j in site S_2, the edge $P_i \rightarrow P_j$ will be inserted in local wait-for graph of S_2.

An example of Centralized approach:

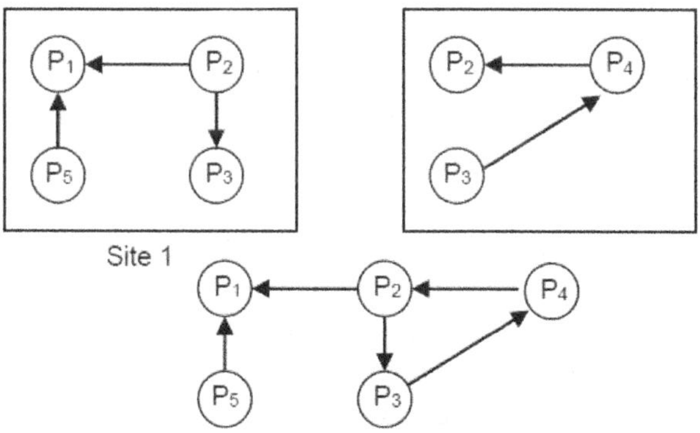

Global Wait for Graph

Note: If any local Wait for Graph has cycle, deadlock has occurred. On the other hand, if there is no cycle in local Wait for Graph, it does not mean that there are no deadlocks as shown above.

Pantham Deadlock: The pantham deadlock exists when the false cycle is detected by control site. Consider the snapshot of system depicted in following figure:

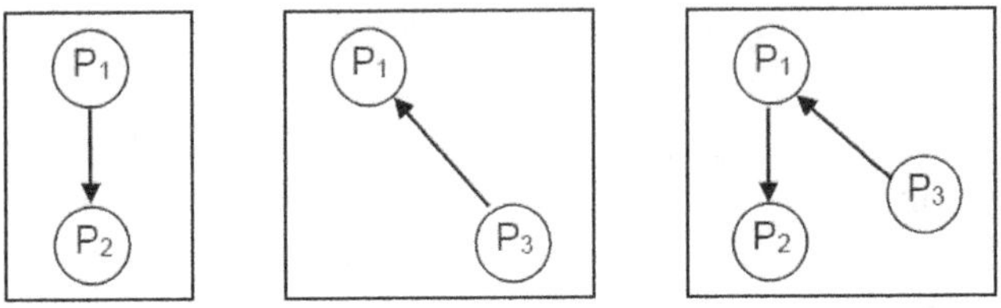

Site 1 Site 2 Global state at control site

If P_2 releases the resource it is holding in site 1, resulting the deletion of edge $P_1 \rightarrow P_2$ in S_1. Then process request resource held by P_3 at S_2 resulting the addition of edge $P_2 \rightarrow P_3$ in S_2. If {insert $P_2 \rightarrow P_3$} from S_2, arrives before the {delete $P_1 \rightarrow P_2$} from S_1, the control site may discover a false cycle $P_1 \rightarrow P_2 \rightarrow P_3 \rightarrow P_1$ after the insert (but before delete). The deadlock recovery may be initiated although no deadlock has occurred.

Drawbacks of Centralized Approach

- Possibility of pantham deadlock. So recovery is initiated which may abort or roll back some processes although there is no deadlock.
- This approach has single point of failure, i.e., if control site fails no deadlock will be detected in system.
- Communication links near control site are likely to be congested because control site receives WFG information from all other sites.

b. Distributed Control: In this approach the responsibility for detecting a global deadlock is shared equally among all sites. The global state graph is spread over many sites and several sites participate in detection of global cycle/knot.

Distributed deadlock detection algorithms are classified as:

- Path pushing algorithms
- Edge chasing algorithms.

Path pushing algorithms: In path pushing algorithm, information about the wait for dependency is propagated in form of paths.

Each site maintains its local WFG. The local WFG in this approach is different from local WFG discusses in centralized approach.

In this approach local WFG includes an additional node P_{ex}

- $P_i \rightarrow P_{ex}$ exists in graph if P_i is waiting for resource residing at another site and held by any process.
- Edge $P_{ex} \rightarrow P_j$ exists in graph if a process at another site is waiting to acquire resource currently being held by P_j in this local site.

Algorithms:

If local Wait for Graph contains a cycle that does not involve node P_{ex} then system is in deadlock state.

If cycle involving P_{ex} exists, then this implies the possibility of deadlock. Suppose at site S_i, local WFG contains cycle involving P_{ex} as

$P_{ex} \rightarrow P_{k1} \rightarrow P_{k2} \rightarrow P_{k3}$ ...... $P_{kn} \rightarrow P_{ex}$

This indicates that P_{kn} in site S_i is waiting for resource located in some other site – say S_j. On discovering this cycle, site S_i sends to S_j, a deadlock-detection message containing information about that cycle.

- When site S_j receives the deadlock detection message, it updates its local WFG with new information. Then it searches newly constructed WFG for cycle not including P_{ex}. If one exists, the deadlock is found and an appropriate recovery scheme is invoked. If cycle involving P_{ex} is discovered, then site S_j transmits a deadlock detection message to appropriate site – say S_k.
- S_k, in return repeats the same procedure. Thus after finite number of rounds either deadlock is detected or deadlock-detection computation halts.

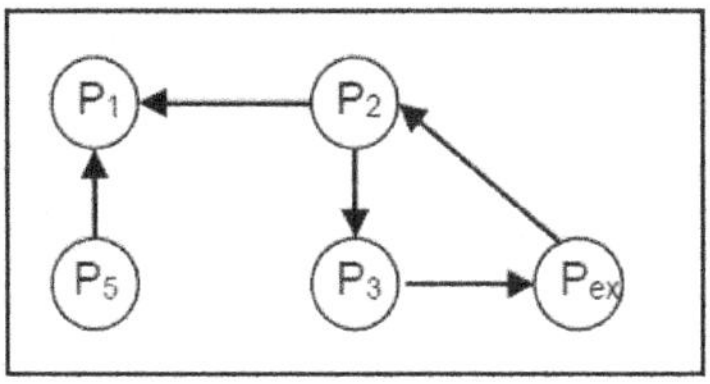

Site 1

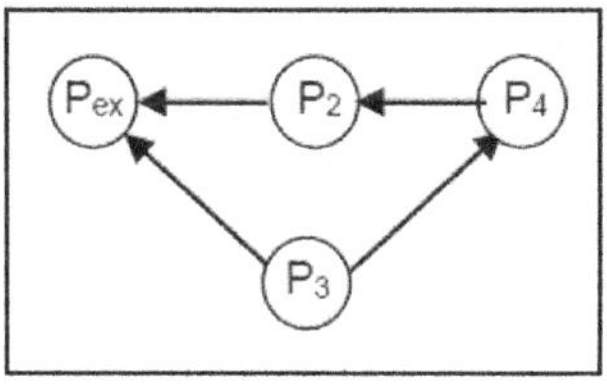

Site 2

Suppose S_1 discovers cycle

$P_{ex} \rightarrow P_2 \rightarrow P_3 \rightarrow P_{ex}$

Since P_3 is waiting to acquire a resource of site S_2, so this cyclic path is transmitted from S_1 to site S_2. When S_2 receives this message it updates its local Wait for Graph as follows:

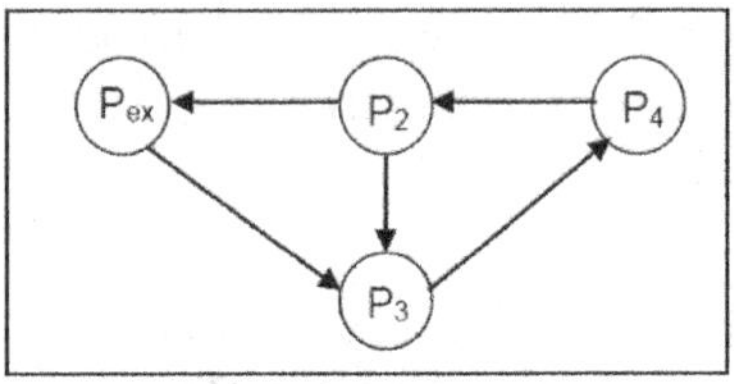

Updated Wait for Graph at site S_2

This Wait for Graph contains cycle ($P_2 \rightarrow P_3 \rightarrow P_4 \rightarrow P_2$) excluding P_{ex} so deadlock is detected.

Note: The outcome would be the same if S_2 detects cycle, first in its local Wait for Graph and sends deadlock–detection message to S_1. In worst case both site may discover the cycle at about same time and two deadlock-detection messages are sent: one by S_1 to S_2 and another by S_2 to S_1. This situation results in unnecessary message transfer and overhead in updating two local Wait for Graphs and detecting cycle in both graphs.

To overcome this problem; the algorithm is updated as:

If site S_k discovers that is local Wait for Graph contains a cycle involving P_{ex} of form

$P_{ex} \rightarrow P_{k1} \rightarrow P_{k2}$ ...... $P_{kn} \rightarrow P_{ex}$

It sends deadlock-detection message to another site only if:

ID (P_{kn}) < ID (P_{k1})

i.e., identification number of P_{kn} is less than that of P_{k1}.

Edge Chasing Algorithm: In edge chasing algorithms, special messages called the probes are circulated along the edges of Wait for Graph to detect the cycle. When a blocked process receives the probe, it propagates the probe along its outgoing edges in Wait for Graph. A process declares deadlock if it receives a probe initiated by itself.

A probe is a triplet (i, j, k) denoting that it belongs to a deadlock detection initiated by P_i, and it being sent by home site of P_j to home site of P_k.

Data structure used in this algorithm:

- A process P_j is said dependent on another process P_k if there exist a sequence P_j, P_{i1}, P_{i2}, ... P_{im}, P_k such that each process except P_k is blocked and each process except the P_j holds the resource for which the previous process in sequence is waiting.

- P_j is locally dependent on P_k if P_j is dependent on P_k and both are at same site.
- A Boolean array dependent$_i$ for each process P_i where dependent$_i$ (j) is true if P_i knows that P_j is dependent on it.

Algorithm:

Initialize dependent$_i$ (j) = false

$\forall_I, \forall_j$

To determine blocked process is deadlocked, following algorithms is executed:

If P_i is locally dependent on itself
  then declare deadlock
else for all P_j and P_k such that

- P_i is locally dependent upon P_j and
- P_i is waiting on P_k, and
- P_j and P_k are on different sites

  send probe (i, j, k) to home site of P_k

end

On receipt of probe, the site takes following action:

if (P_k is blocked && dependent$_k$ (i) = false && P_k has not replied to all requests of P_j)

then
  dependent$_k$ (i) = true
  if (k = i)
      declare P_i is deadlocked
else for all P_m and P_n such that

- P_k is locally dependent upon P_m
- P_m is waiting on P_n
- P_m and P_n are on different sites

  send probe (i, m, n) to home site of P_n

end

Example:

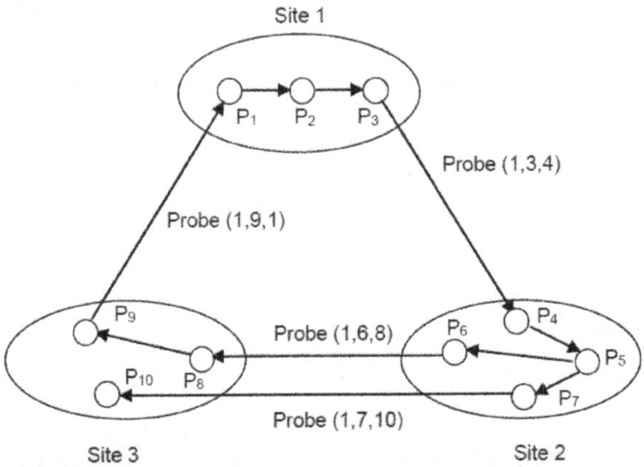

c. Hierarchical Deadlock Detection Approach: In this approach, the sites in the system are arranged in a hierarchical manner i.e., using inverted tree like structure. A site is capable of detecting the deadlock in all its children sites.

Various algorithms were proposed using this approach.

The Menasce-Muntz Algorithm: In this approach, sites are arranged in tree like structure. Each site works as controller for its all descendent nodes. The controller at bottom most level manages only its resources and is known as leaf. Thus a leaf just maintains the part of global WFG concerned with allocation of its local resources. All the non leaf sites collect the Wait for Graph of all its child sites and constructs the WFG using all information, and detects the cycle.

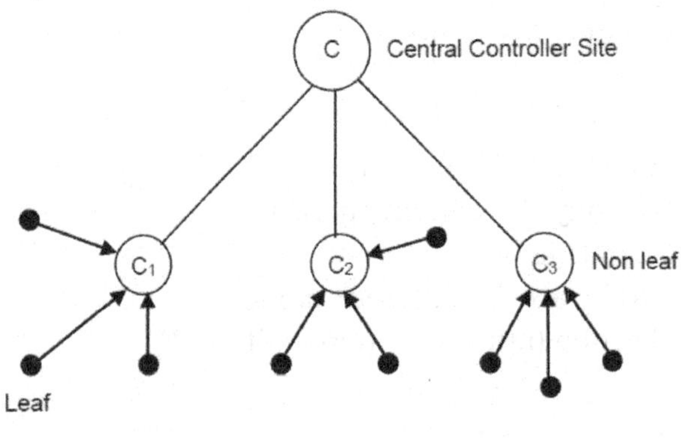

3.2 AGREEMENT PROTOCOLS

3.2.1 Prerequisites

Assumptions about distributed system:

Construction : Collection of entities spatially separated and communicating only by exchanging messages.

Purpose : Resource sharing.

Don't have : — shared clock and memory
— view of entire correct state.

Haven't (yet) said anything here about :
—trust
—identity.

Trust is measure of willingness to proceed with an action (decision). Without risk, either parties refuse to interact or require several restrictions and complex (costly) control.

Trust is required in order for multiple entities to cooperate and share resources to achieve some application values.

Identity is used in distributed system to represent distinct entities.

Thus in distributed system, where several sites are working together, communicating via messages, some type of trust between them is required. To develop trust among them, sites make some sort of agreements for mutual working. The necessary condition for agreement maintenance is that every site should be aware of other sites and their processing. Every site should know the input taken and output produced by other sites.

When the system is free from failure, an agreement can easily be reached among different sites. For example, sites can reach an agreement by communicating their values to each other and then by taking a majority vote or a minimum, maximum, mean etc. of those values. However, when the system is prone to failure, the trust factor lessens down because faulty site may mislead to other sites. It can send conflicting values to other processors preventing them from reaching an agreement.

So the entire process of reaching an agreement in presence of faulty sites, us known as agreement protocol and it involves several round of message exchange among sites.

Agreement Problem

Nonfaulty processors (sites) should be able to reach common agreement, even if certain components in system are faulty.

3.2.2 Model of Processor Failures

A model of failure specifies in what way a faulty process can behave:

Crash failure	:	A faulty processor stops prematurely and does nothing from that point on Before stopping, however, it behaves correctly.
Send omission	:	Processor omits to send messages to same processors. For example, a processor is supposed to broadcast the message to all other processors, but it sends the message to only few processors.
Receive omission	:	Processor, omits to receive message sent to it.
Arbitrary failure	:	(Also known as Byzantine or Malicious failure) Processor can exhibit any behavior whatsoever. For example, it can change state arbitrarily or it may send fictious messages to other processors to confuse them.

3.2.3 Classification of Messages

We know that the distributed system works on messages exchange and their processing. There are two types of messages:

- **Authenticated Messages:** Authenticated messages are not allowed to change by the faulty processor. The faulty processor only can receive such messages and verify its authenticity and then transfer it to others. Due to this feature authenticated messages are also called 'Sined Messages'.

- **Non-Authenticated Messages:** In the case of non-authenticated messages, the faulty processor can change the content of message before it relays the message to other processors. A processor has no way to verifying the authenticity of a received message. A non-authenticated message is also known as 'Oral Message'

Note: It is easier to reach agreement in an authenticated message system because faulty processors are capable of doing less damage.

3.2.4 Classification of Computation

A distributed computation can be classified as:

- **Synchronous:** The synchronous computation model specifies that the process in the system execute step by step. The process receives a message for computing task, perform the computation task and then forward the message to any other process for further computation. In this way, the computation task is performed step by step and one step is known as round. The messages are received and processed one after another. If there is any delay between messages then all computation work will slowdown and decrease the performance of system.
- **Asynchronous:** In asynchronous computation model, it is not necessary to perform the computation task in steps. The computation can be performed at any time without any sequence of steps.

3.2.5 Classifications of Agreement Problem

Agreement problem in distributed system arise due to nature of system. The requirement of agreement-distributed problems is that all non-faulty processors must reach a common agreement. The well-known agreement problems in distributed system are:

- Byzantine Agreement Problem.
- Consensus Problem
- Interactive Consistency Problem.

Byzantine Agreement Problem (BAP): BAP is generally also known as Byzantine Generals Problem. The problem can be described as follows:

"In a battle, there are several divisions of army, each commanded by its own Lieutenant. All Lieutenants are headed by a General. The general and lieutenant must reach agreement on whether or not to attack. It is crucial that all lieutenants and General should agree, since on attack by only some of divisions would result in defeat. The various divisions are geographically dispersed and the General and lieutenants cam communicate with one another via messengers who run from camp to camp."

They may not be able to reach agreement for at least one of the following reasons:

- The messenger is caught by enemy and thus may be unable to deliver their message.
- The general may be traitor, preventing the loyal lieutenants from reaching agreement.
- The lieutenant may be traitor.

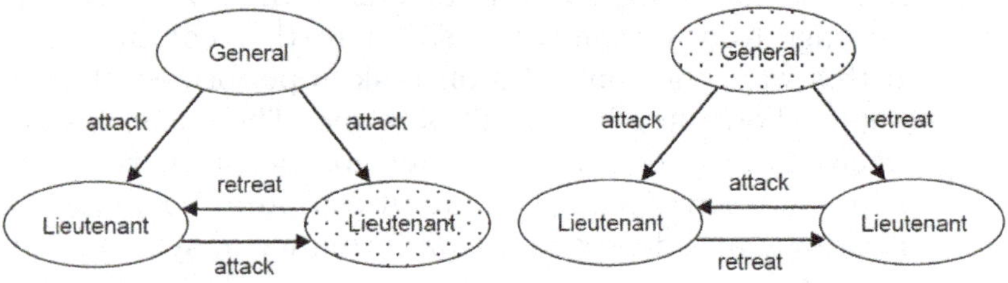

One lieutenant is faulty (is traitor)

General is faulty (is traitor)

In other words the Byzantine agreement problem can be described as:

"An arbitrarily chosen processor, known as source processor (also known as commander) broadcasts its initial value to all other processors." The solution to BAP should follow following objectives:

- All non-faulty processors should agree on same value.
- If commander is not faulty, the commonly agreed value should be the value, sent by commander.
- If commander is faulty, then all non-faulty processors can agree on any arbitrary value.

Consensus Problem: The consensus problem differs from Byzantine generals problem. In the Consensus all processors proposes a value, and in Byzantine a distinguished processor (source processor) supplies a value and the others are to agree upon that.

As we know consensus means "a generally accepted opinion or decision among a group of people"

The distributed consensus problem arises when several processes need to reach a common decision (depending on their initial inputs) despite failures. The importance of this problem is due to its omnipresence in distributed computations we need consensus to implement reliable communication, atomic commitment, consistency checks, resource allocations etc.

Let us consider a system in which consensus is needed. As each processor has its initial value, so these initial values are broadcasted to others. For reaching consensus, all processors follow a protocol, which satisfies following properties:

1. **Termination:** Every processor decides some value.
2. **Agreement:** No two non-faulty processors decide differently.
3. **Uniform Integrity:** Every processor decides at most once.
4. **Uniform Validity:** If a process decides v, then v was proposed by some processor.

Means, all non-faulty processors must agree on common single value. If the initial value of all non-faulty processors are same, then they agree on that value otherwise they agree on majority value.

Interactive Consistency Problem: Interactive consistency problem is another variant of consensus, in which every processor proposes a value and processors agree on a vector of values, one for each processor. Generally we call this 'decision vector'. The protocol for interactive consistency problem should meet following conditions:

- **Agreement:** All non-faulty processor agree on same vector
- **Validity:** If the ith processor is non-faulty and its initial value is v_i, then i^{th} value to be agreed on by all non-faulty processors must be v_i.

If j^{th} processor is faulty, all non-faulty processors can agree on any common value for v_j.

Note: In Byzantine problem, unsigned messages are used.

3.2.6 Solution to Byzantine Agreement Problem

We know that in Byzantine agreement problem, an arbitrarily chosen processor (called source processor) broadcasts its initial value to all other processors.

Any solution to Byzantine agreement problem must guarantee following rules:

- **R1:** All non-faulty processor should agree on same value.
- **R2:** If source processor is non-faulty, the common agreed upon value should be the initial value of source.

This it is obvious that processors exchange the values to reach consensus. The faulty processors may confuse other processors by sending conflicting values.

Upper Bound on Number of Faulty Processors in Byzantine Agreement Problem Solution

The number of faulty processor plays an important role for finding solution to BAP. As faulty processor have a great impact on non-faulty processors, so major requirement for getting an agreement value is that non-faulty processors should not be influenced by presence of faulty processors. Therefore we can say number of faulty processors should not be greater than that of non-faulty processors.

There have been many researches done for finding the limited number of faulty processors in the system. One such research performed by Pease et al shows that number of faulty processors should not be larger than ((n-1)/3) where n is number of non-faulty processors.

$$m \ngtr \frac{n-1}{3}$$

Lamport proposed the first solution for Byzantine agreement, which requires m +1 rounds of message transfer where m is maximum number of faulty processors.

Treatment of Impossible Result: Sometimes, the agreement problem may lead to such a situation, which is quite impossible to solve. The situation where agreement is impossible is called impossible result situation.

Let us check the impossible result situation in a system with three processors among them one is faulty.

Case 1: Source Processor is not faulty.

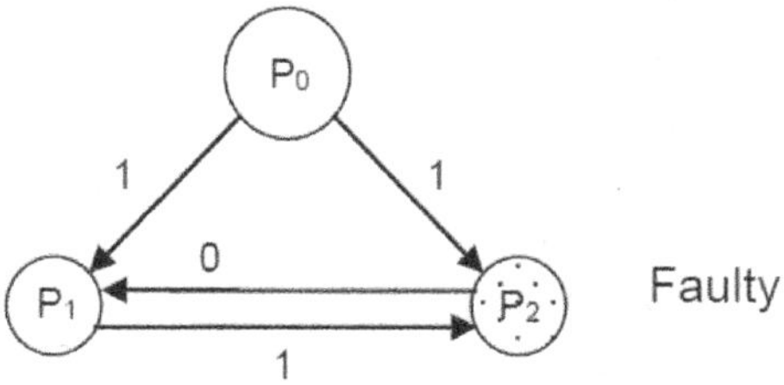

Assume P_2 is faulty. Although P_1 is receiving two conflicting values from P_0 and P_2, it must accept 1 as agreed upon value to satisfy rule 1 (R_1) i.e., agreed upon value sent by source.

Case 2: Source processor is faulty.

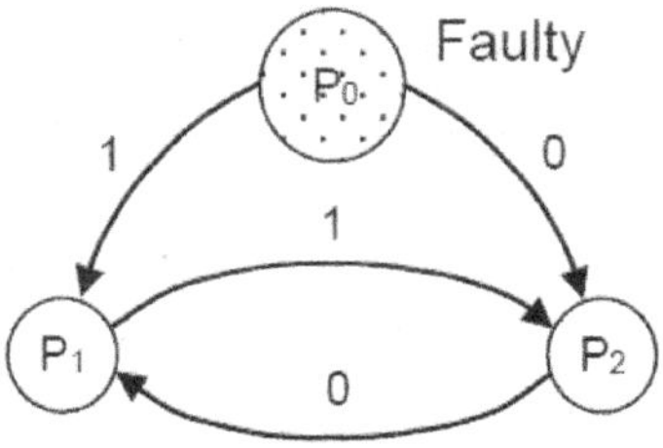

However if this is followed in case 1, P_1, will agree on 1 and P_2 will agree on 0, which will violate rule 2 (R_2).

Therefore no solution exists for Byzantine agreement problem for three processors, which will work under single processor failure.

Lamport Algorithm for Byzantine Agreement Solution

Lamport et al proposed an algorithm for solving the Byzantine agreement problem. This algorithm is also called "Lamport-Shestak-Pease-Algorithm. They proposed the algorithm for oral messages (Non-authenticated messages), so it is also referred as "Oral Message Algorithm (OM)."

This algorithm works when following is true.

$n \geq 3m + 1$

Where n = total number of processors.

m = number of faulty processors.

The algorithm works differently with faulty processors or without faulty processors.

OM (o): Algorithm (i.e., m = 0)

Step 1: Source processor sends its initial value to each and every processor.

Step 2: Each processor uses this received value (If no value is received it uses a default value 0).

OM (m) Algorithm (i.e., m > 0)

Step 1: The source processor sends its value to every processor.

Step 2: If a processor does not receive any value it uses a default value of zero.

Step 3: If a processor (P_i) receives a value v_i from source, then it behaves like source processor, and initiates OM (m – 1) algorithm, where it broadcasts this value v_i to (n – 2) processor (i.e., excluding itself and the source from where it is receiving value).

Step 4: If for a processor P_i, v_j, $(j \neq i)$ is value received from P_j, the P_i uses the majority value as agreement value.

Example 1: Consider a system with four processors P_0, P_1, P_2, P_3 among them one is faulty.

Case 1: P_0 executes the algorithm OM (1) and suppose P_2 is faulty.

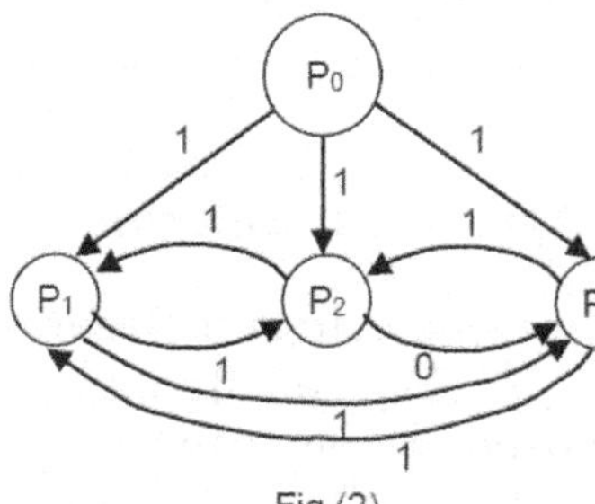

Fig (2)
P_1, P_2, P_3, executes o m (0)

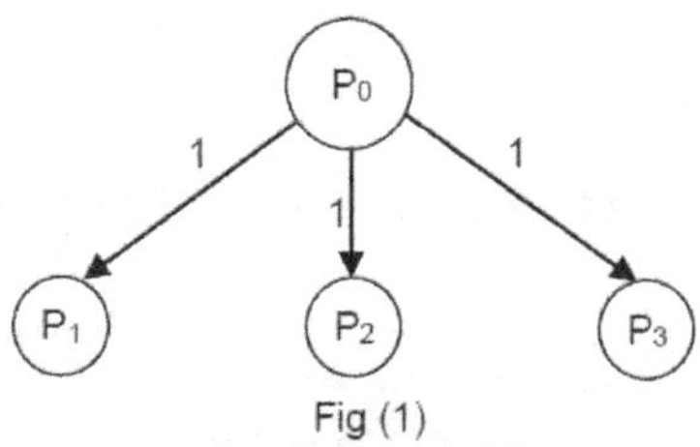

Fig (1)
P_0 initiates o m (1)

The majority value for Byzantine solution:

Processor	Receive Values	Common majority values
P_1	(1, 1, 1)	1
P_2	(1, 1, 1)	1
P_3	(1, 1, 0)	1

Case 2: P_0 executes the algorithm OM (1) and it is faulty.

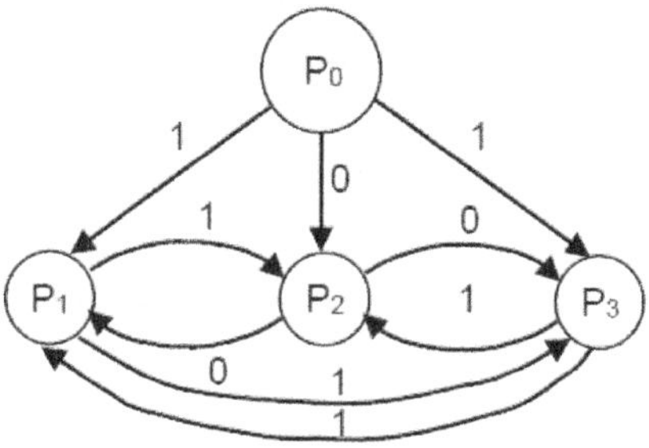

The majority value for Byzantine Solution:

Processor	Receive Values	Common majority values
P_1	(1, 0, 1)	1
P_2	(0, 1, 1)	1
P_3	(1, 1, 0)	1

3.2.7 Applications of Agreement Protocols

The agreement problem is very much found in the situation where multiple processors work together. So these algorithms are helpful for finding their solution. These algorithms play an important role in distributed systems such as clock synchronization and atomic commitment. Let us have a brief look on these applications:

- Clock synchronization
- Atomic Commitment in Distributed Database system

Clock Synchronization

In the distributed system, various sites work together and having apart of processing. So, these sites need to maintain the physical clock for synchronizing with each other. He physical clocks must be periodically synchronized. But the re-synchronization becomes very difficult if there is a case like that of Byzantine agreement problem.

So it is needed to take an agreement value for all the clocks of all the sites. In this situation the agreement problem algorithms play an important role and help to find a common agreement value. Sum assumptions which are required for fault tolerant clock synchronization are:

- Initially, all the clocks are synchronized o same value.
- A non faulty process's clock runs approximately at correct rate.
- The non faulty process can read the clock value of other non faulty processes and there may be a very small error in reading.
- There is no assumption for faulty clock.

So the agreement algorithm should be modified for clock synchronization. Hence is known as ***Clock Synchronization Algorithm***. This algorithm should satisfy the following conditions for finding the agreement solution:

- At any time, the value of clocks of all non faulty processors must be approximately equal.
- There is small bound on the amount by which the clock of non faulty processors is changed during each resynchronization.

This means the clock value does not change randomly during synchronization. In this manner, the clock time will not have a large change from that of a real time.

The Following two types of clock synchronization algorithms were given for the Fault Tolerance with the situation of Byzantine Problem:

1. **Interactive Convergence Algorithm:** It is called Interactive Convergence Algorithm because It causes the convergence of non faulty clocks. According to this algorithm, all the clocks are synchronized initially. They are synchronized but there is made a bound for synchronization that the two non faulty clocks never differ by the value 'd'

Algorithm: First of all, each processor reads the value of clocks of other processors and sets its clock value to the average of these values. However, if a clock value differs from is own clock value b more than 'd' it replaces that value by its own clock while taking the average.

2. **Interactive Consistency Algorithm:** It is called Interactive Consistency Algorithm because all non faulty processors get the consistent views of all the processors clock. The concept of this algorithm is to take the median of the clock values. The calculation of median of the clock values is useful for better estimation of clock values. It also handles the problem where the different clock values to different processors received. In this situation, the two processors can calculate the same median, if they get the same set of clock values.

   It means, one condition should be satisfied be satisfied by all processors, that is, "*Any two processors obtain approximately the same value for a processor P's clock whether the P is faulty or not*". All the processors compute the same value and their values should also be close to values of faulty processors. In this situation, one another condition is must to satisfy, that is "*every non faulty processor obtains approximate correct value for the clock of Q if Q is non faulty processor*".

   Hence, it is true if most of the processors are non faulty, the median of their clock value is equal to correct clock value or lies between two correct values.

Atomic Commitment in Distributed Database System

In the case of distributed database, all he sites must agree for a transaction either to ***commit*** it or to ***abort*** it. Atomic Commitment in Distributed Database System comprises two phases:

- *Phase 1:* Every site executes their part of distributed transaction and broadcast their decisions (Commit or Abort) to all other sites.
- *Phase 2:* Each site, decides whether to commit or abort its part of distributed transaction, based on what it has received from other sites in first phase.

Since every site receives an identical response from all other sites, they will reach the same decision. But if some sites behave maliciously, they can send a conflicting response to other sites, causing them to make conflicting decisions. In this situation, we can use Byzantine Agreement to ensure that all non faulty processors reach a common decision about a distributed transaction. It works as follow: *In the first phase, after a site has made a decision, it starts the Byzantine Agreement. In the second phase, processors determine a common decision based on agreed vector of values.*

EXERCISE

Q1. Write short note on following:

a. Deadlock Detection

b. Deadlock Avoidance

c. Deadlock Prevention

Q2. Differentiate the centralized, distributed and Hierarchical control organization for distributed deadlock detection. Also discuss the Edge-Chasing Algorithm for detecting deadlock in distributed environment.

Q3. Why the deadlock prevention and deadlock avoidance approaches are not efficient in distributed environment and deadlock detection approach is only useful? Also differentiate the centralized, distributed and Hierarchical control organization for distributed deadlock detection.

Q4. What are the Phantom Deadlocks? How the One phase algorithm removes the possibility of false deadlock detection?

Q5. Discuss how the Path Pushing Algorithm detects the deadlock in distributed environment?

Q6. Discuss Briefly about the Agreement Problem.

Q7. How can the fault tolerance is achieved by Byzantine solution?

Q8. Classify the different types of processor failures. What is Byzantine Agreement Problem? What is the upper bound on number of faulty processors in solution of Byzanine Agreement problem? Discuss the Lamport-Shostak-Pease Algorithm for solving this problem.

Q9. What is the clock synchronization algorithm?

Q10. Discuss the atomic commitment in distributed database system.

Q11. Differentiate the following:

a) Communication & Resource Deadlock

b) Asynchronous & Synchronous computation

c) Authenticated and Non-Authenticated messages

4

Advance Issues in Distributed System

4.1 DISTRIBUTED FILE SYSTEM

4.1.1 Introduction

File systems are responsible for organization, storage, retrieval, naming, sharing and protection of files.

In the distributed system, the file system works as resource management component, which manages the availability of files among various sites. Thus, it is a common file system, shared by all computers in systems. The two main goals of DFS are:

- **Network Transparency:** Means availability of files distributed over the network in same way the file residing at local site.
- **High Availability:** The users in distributed network can easily access the file irrespective of their physical location.

File Sever: The file servers are the specialized servers dedicated to only file management in distributed system. Other machines request for the remote file to the file servers. The responsibility of file server is to search the location of requested file and then make it available to the requesting site. Thus, file servers are always dedicated to request fulfillment of all other machines (clients). The clients may have their local storage disk or not. The component of file servers are:

- **Name Server:** The name server maps the logical name of requested file into exact physical address. The process of determining physical address of logical object is known as name resolution.
- **Cache Manager:** The cache manager is a process that implements file caching. In file caching, a copy of data stored at remote file server is brought to the client's machine when referenced by client. The subsequent access to data are performed locally at the client, thereby reducing the network access delay. The cache manager can be present at clients and server.

Architecture of Distributed File System is depicted in figure 4.1.

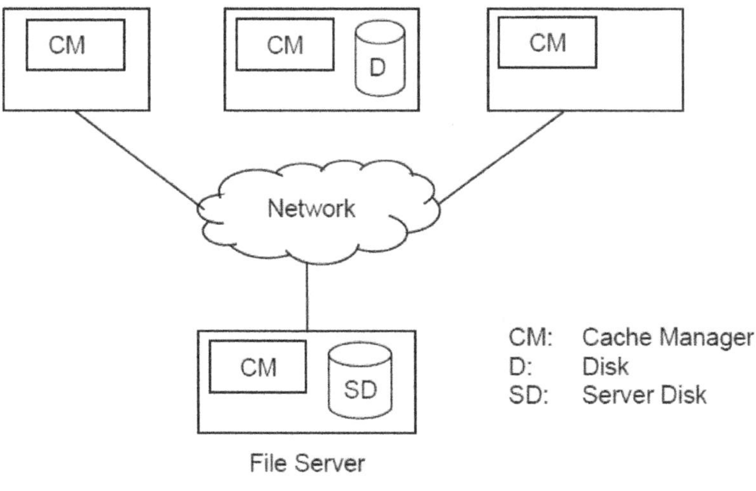

Figure 4.1: The Architecture of DFS

Typical Data Access in Distributed File System is illustrated in figure 4.2.

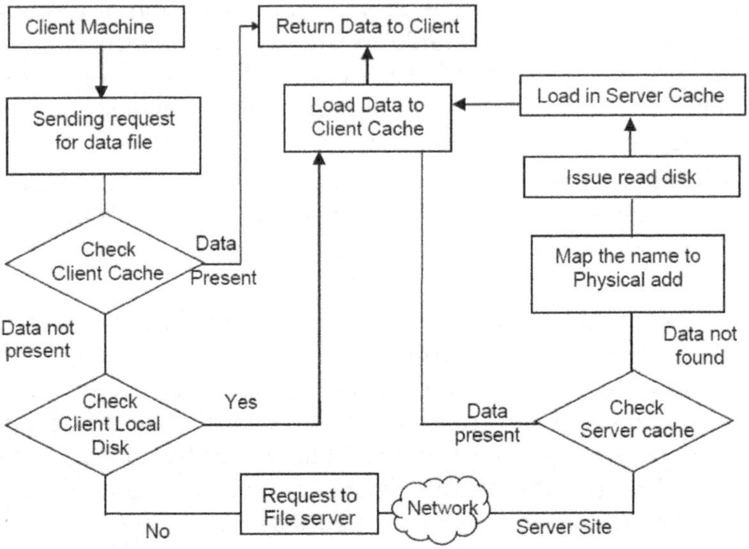

Figure 4.2: Data Access in Distributed File System

4.1.2 Mechanism for Building Distributed File System

The common mechanisms for building up the distributed file system are as follows:

1. Mounting: Mounting a file system means assigning the root directory of the new file system to a subdirectory of root directory. After this new file system is accessible to the user as a sub directory of root.

Thus, mounting mechanism allow the binding of different filename spaces together. This combination forms a hierarchical structured tree.

Note:

- The namespace is the collection of files on disk and assigning a name to that space for future use.
- The node in namespace tree on which the mounting is done is called Mound point.

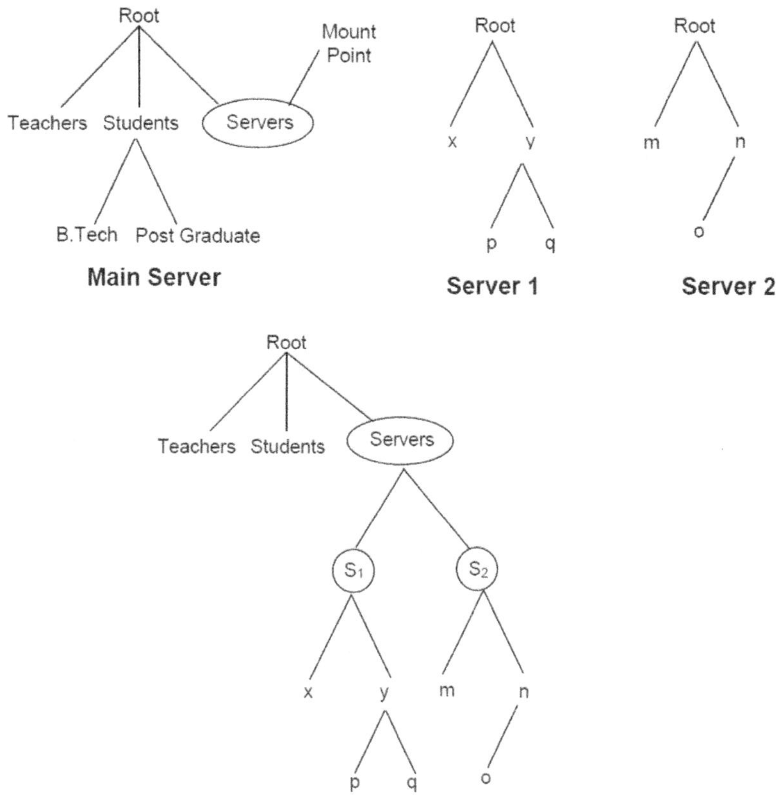

Figure 4.3: Server 1 and Server 2 Mounted on Main Server

2. Caching: Caching is most common mechanism applied during the design of the distributed file system. In file caching, a copy of data stored at a remote file server's cache is brought to the client's cache, when referenced by client. Subsequent access to the data is performed locally at the client, thereby reducing network access delay.

The file system performance can be improved by caching since accessing remote disks is much slower than accessing local memory.

3. Hints: Caching improves file system performance by reducing the delay in accessing data. However when multiple clients cache and modify shared data, the problem of cache consistency arises.

It must be guaranteed that cached data is valid and copy of data recently updated in another client's cache or in file server's cache does not exist.

To guarantee this, the concept of hints is used.

A times stamp based method is used to validate cached blocks before they are used.

Each data item in cache is tagged with hints. The hints for cached data object is also stored in cache of the system and used again at the time of next data access. The whole concept of hints is as follows:

- When a cached entry is modified, the new value is kept in the cache, but is also immediately sent to server. As a consequence, when another file reads the file, it gets the most recent value. Each data item in cache is tagged with hint, which is the time when the block was last modified at server.
- If the hint value (last modification time) of the cached version and server version is same, the client's cache is valid and up-to-date. If they differ cached data has been updated at server and cached entry is invalidated, resulting in request to server for relevant data.

4. Bulk Data Transfer: The transfer of data is not an easy task in network. The data transfer may be expensive due to use of various protocols. These protocols involve splitting the data into packets at sender site and assembling the packets again at receiver site. Acknowledgements are also sent for each received packets by receiver. So actual transit time across a local area network can be insignificant.

Transferring data in bulk reduces the protocol processing over-head at both servers and clients. In bulk data transfer multiple consecutive data blocks are transferred from servers to clients instead of just the block referenced by clients.

The bulk transfer of data reduces the access of remote file because multiple data blocks are accessible in single seek.

5. Encryption: Encryption is the process for data security in distributed system. In the distributed system, as data is available to all sites, so some encryption techniques must be used.

The message is transferred with the help of a key. The two systems in the distributed network setup the key for communication between them. In the distributed environment, the text is always transferred from one machine to another machine in the form of encrypted text only. The plain text can never be transmitted from one system to another.

4.1.3 Design Issues for Distributed File System

The design of DFS depends on various facts. The knowledge of these various important issues help someone to understand the DFS in letter way. Let us discuss these issues:

1. Naming and Name Resolution: Naming is mapping between logical and physical objects. For instance, users dead with logical data objects represented by file names, whereas the system manipulates physical blocks of data, stored on disk tracks. Usually, a user refers to a file by taxual name. That name is mapped to a lower level numerical identifier than in turn is mapped to disk blocks.

This multilevel mapping provides users with an abstraction of a file that hides the details of how and where on disk the file is stored.

In DFS, a new dimension is added, that is, hiding where in network the file is located. In conventional FS, the range of mapping is an address within a disk.

In DFS, this range is argumented to include specific machine on whose disk file is stored.

The two notions regarding name mapping in DFS are as follows:

i. **Location Transparency:** Means path name gives no hint as to where data is allocated. The pathname server 1\dir 1\x tells every one that x is located on server 1. But it does not tell where the server is located. The server is free to move anywhere in network, without path name having to be changed.

However, suppose file x is extremely large and space is tight on server 1, the system cannot move file to other server automatically because it changes the path. If it happens programs that have previous path built into them will cease to work.

ii. **Location Independence:** The naming scheme is location independent if the name of file in application programs need not to be changed when the file physical storage location changes.

Naming and name resolution is provided by name servers. In the distributed system, name servers are responsible for name resolution.

The approaches to name resolution in distributed environment are as follows:

- All clients send their name resolution queries to single name server, which maps name to objects. This approach has drawback, if the name server crashes, the entire system is affected.
- The second approach involves having several name servers wherein each server is responsible for mapping objects stored in different domains. When a name is to be mapped to an object, the local name server is queried. The local name server may point to remote name server for further mapping of the name

2. Write Policy: Writing policy decides when a modified cache block should be transferred from client to server. Some write polices are as follows:

i. **Write – trough:** According to this policy , the requests made by the clients are carried to the server immediately. In this way, the server keeps entire information regarding the client requests and their fulfillment. This policy is important to us, if any client clash occurs. If a client system crashes, the information of client requests remains on the server also so the data loss is very small.

ii. **Delayed writing:** This policy is concerned with the delay of writing server. When the data is modified, the modifications are reflected to server after a time delay. The data blocks are modified and kept in local center of client and after many writes this block is transferred to server. The intermediate results are not stored to server because it is deleted from local cache in very short time.

    The nature of block modified should be known and the length of delay period is also to be found. But this writing policy is not so reliable because the data remains present locally on the client cache for some time. If any client gets crashed, the data present in local cache is also lost.

iii. **Write on close policy:** This policy delays the updating of files on the server till the files are not closed. at the client side. It means, client updates the files locally and then closes it. Then the server updates the file.

4.2 DISTRIBUTED SHARED MEMORY

4.2.1 Introduction

Distributed shared memory (DSM) is an abstraction used for sharing data, between computers that do not share physical memory.

Traditionally, distributed computing has been based on the message passing model in which processes interact and share data with each other by exchanging data in form of messages. The shared memory model provides a virtual address space that is shared among all node in different distributed system (Figure 4.4)

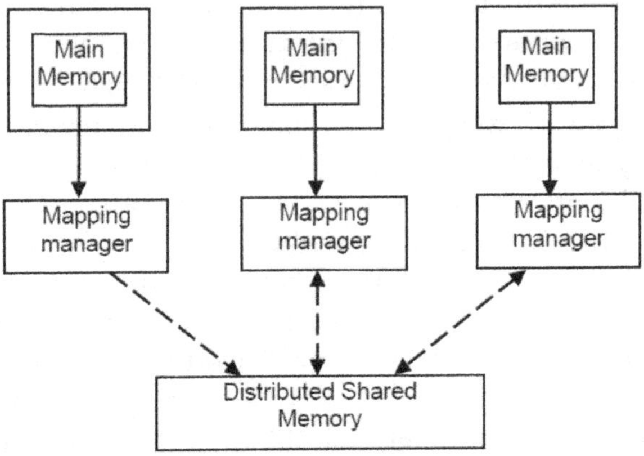

Figure 4.4: Distributed shared memory

In the systems that support distributed shared money, data moves between main memories of different nodes.

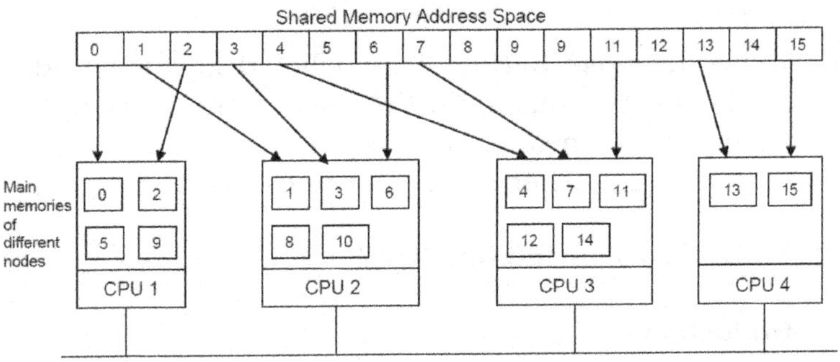

Figure 4.5: Pages of Address Space Spread Over Four Machines

The mapping manager is the layer of software embedded with OS that maps the shared memory address to the physical memory.

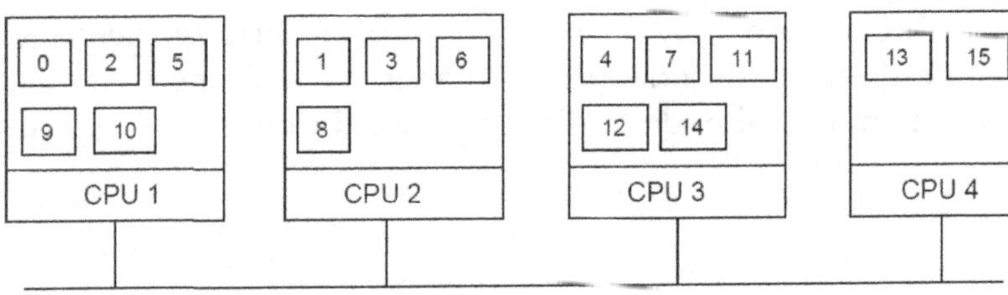

Figure 4.6: Situation after CPU 1 references Page 10

4.2.2 Message Passing vs Distributed Shared Memory

1. In message passing model, program make shared data available through explicit message passing. So programmers need to be conscious of data movement between processes. The programmers have to explicitly use the communication primitives (such as SEND and RECEIVE).
2. In contrast, DSM systems hide this explicit data movement and provide a simpler abstraction for sharing data.
3. In message passing model, the data moves between two different address spaces. So passing data structures containing pointers is generally difficult and expensive.
4. In message passing model, specific piece of data is transferred but in the systems with DSM, entire block or page containing the data is transferred, so DSM exploits the concept of locality of reference exhibited by programs.
5. The physical memory available at all the nodes of DSM system combined together is enormous. This large memory can be used to efficiently run programs that require large memory without involving the disk due to swapping in traditional systems.
6. Process communicating via message passing must execute at same time while process communicating via DSM, may execute with non-overlapping lifetime. A process can leave data in an agreed memory location for the other to examine when it runs.

4.2.3 Algorithms for Implementing DSM

While implementing the DSM, the control issues are:

- How to keep track of the location of remote data?
- How to overcome the communication delay?
- How to make shared data concurrently accessible at several nodes in order to improve performance?

The Basic Algorithms for Implementing DSM:

1. Central-Server Algorithm

2. Migration Algorithm
3. Read Replication Algorithm
4. Full Replication Algorithm

1. Central-Server Algorithm: In this approach, central-server maintains all the shared data. It services the read requests form clients by returning the data items to them. It updates the data one write requests by clients and returns acknowledgement messages. Therefore, various clients directly access the memory of server. In this approach, if the data is not found in client's local memory, the request is forwarded to central server, the central server fetches the whole page containing required data, in its local memory and mapping manager maps the address of remote data to the address of server's memory.

Drawback: Central-server is single point of failure.

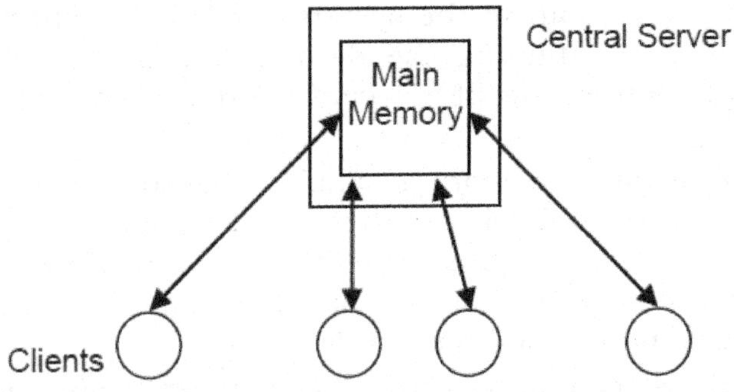

2. The Migration Algorithm: In this approach, data is shipped to the location of data access request, allowing subsequent accesses to data to be performed locally. The migration algorithm allows only one node to access shared data.

As the whole page containing data item migrates, this algorithm takes the advantage of locality of reference exhibited by programs.

Note:

- On memory access fault (memory address is of remote page), a fault handler broadcasts a query to locate the page, and migrate that page in process's address space.

The mapping manager maps the reference of remote page to the process's address space.

- Several processes can share a page at a site.

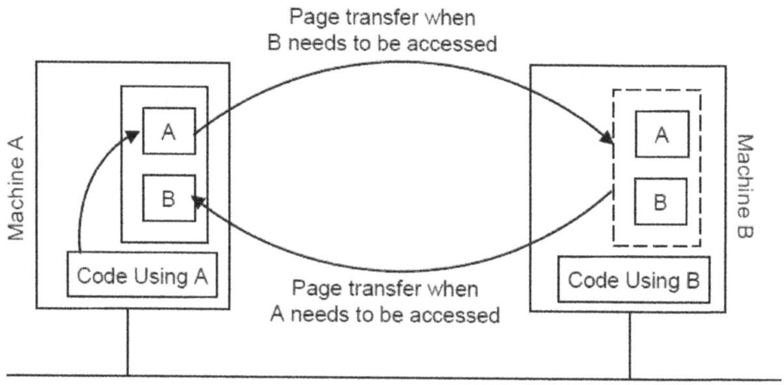

* A and B are two independent data items.

3. The Read Replication Algorithm: The read replication algorithm extends the migration algorithm by replicating data blocks and allowing multiple nodes to have read access or one node to write access. This algorithm improves the system performance by allowing multiple nodes to access data concurrently.

4. The Full Replication Algorithm: It allows multiple nodes to have read and write access to shard data blocks. Because many nodes can write shared data concurrently, the access to shared data must be controlled to maintain consistency.

One simple way to maintain consistency is to use a gap-free sequencer. In this scheme, all nodes wishing to modify shared data will send the modifications to sequencer. The sequencer will assign a sequence number and multicast the modification with sequence number to all the nodes that have a copy of shared data item. Each node processes the modification requests in sequence number order.

4.2.4 Memory Coherence

Although the DSM allows the concurrent access of shared data at many nodes, but these concurrent accesses must be controlled.

Memory is coherent if the value returned by a read operation is correct (the value stored by most recent write operation).

Thus to maintain the coherence of shared data items, a mechanism that controls or synchronizes the accesses is required.

Following are the several forms of memory coherence:

- **Strict Consistency:** The memory is strict consistent if a read returns the most recent write. To maintain strict consistency, as soon as any sites updates the shared data, the replica of whole page must be migrated to all sites that have older version of page. So this approach requires more data movement.
- **Sequential Consistency:** A system is sequentially consistent if the result of any interleaved execution of all the processes is same as they were executed in sequential order and program order of every individual operation is preserved in interleaving.

  Sequentially consistent DSM is implemented by using single server to hold all the shared data and by making all processes perform read and write by sending requests to server. The central server performs the virtual interleaving of all the process's read and write operations against single virtual image of memory, such that program order of every individual process is preserved in this interleaving and each process always reads the latest value written within interleaving.

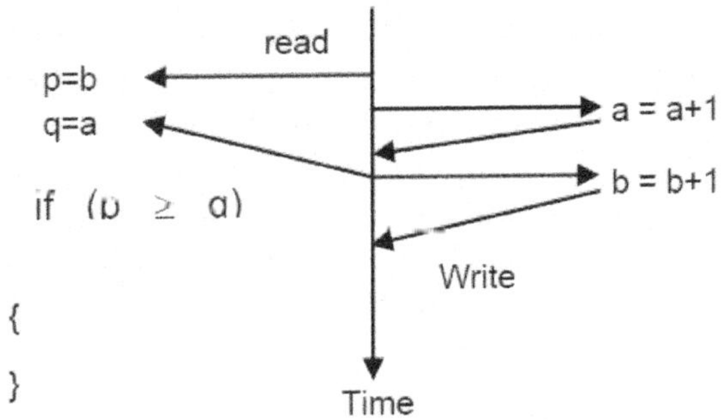

Interleaving Under Sequential Consistency

- **Weak Consistency:** This approach exploits the concept of synchronization operations for accessing shared data, and it leaves the problem of consistency upto programmer. For example, if the programmer uses a lock or semaphore to implement critical section, then DSM system can assume that no other process may access the data items accessed under mutual exclusion, so the execution appears to be sequentially consistent. Thus, by leaving the consistency problem upto programmer, Weak consistency avoids the cost of sequential consistency while retaining the effect of sequential consistency.
- **Release Consistency:** Release Consistency is further relaxation of weak consistency, without a significant loss of coherence. Under the release consistency model, a process need not to block data items when it makes updates within critical section. Also its updates are not propagated until leaves the critical section by releasing the lock. Furthermore, updates are collected and sent in a single message only the final update to each data item need to be sent. For example,

Process 1

```
acquirelock ( )
a = a + 1; b = b + 1
releaselock ( )
```

Process 2

```
acquirelock ( )
print (a, b)
releaselock ( )
```

Under release consistency, if process runs first, it will not block a and b when it access a and b. DSM run time system notes which data is being updated but need take no further action at that time. It is only when process has released the lock, communication is required.

4.2.5 Coherence Protocols

To provide concurrent access, DSM systems make use of data replication, where copies of data are maintained at all the nodes accessing the data. A fundamental problem with data replication is difficulty in ensuring that all copies have same information and nodes do not access stale data.

Two main implementation approach have been derived to maintain the coherence and to propagate updates made by one process to other

- Write Update.
- Write Invalidate.

i. Write Update Protocol: This protocol is also known as multiple-reader/multiple-writer sharing. In this protocol, the updates made by a process are made locally and multicast to all other replica managers possessing a copy of data items, which immediately modify the data read by local processes.

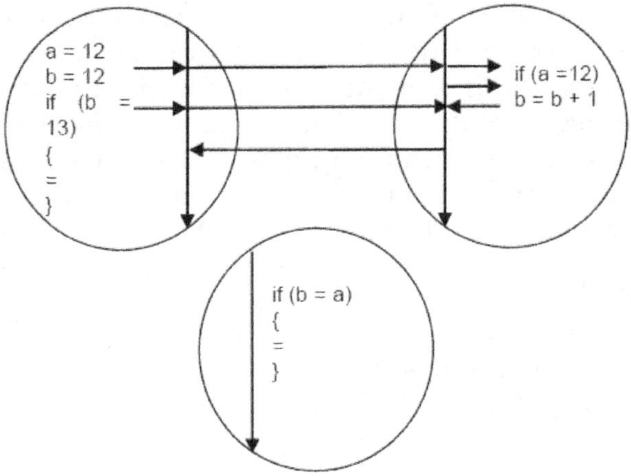

This Protocol achieves the sequential consistency if the multicasting of updation messages is totally ordered.

ii. Write Invalidate: This is also known as multiple-reader/ single-writer sharing. In this protocol, at any time, a data item may either be accessed in read only mode by one or more processes or it may be read and written by only one process. An item that is currently accessed in read only mode can be copied infinitely to other processes. When a process attempts to write to it, multicast message is first sent to all other copies to invalidate them and this is acknowledge before the write can take place. Thus, other processes are prevented from reading stale data. Any process from attempting to access the data item is blocked if a writer exists.

4.3 ADVANCE ISSUES IN DISTRIBUTED SCHEDULING

4.3.1 Distributed Scheduling

As the distributed system consist of a collection of autonomous computers, connected by a local area communication network. The users submit the task at their host computers for processing. The need of load distributing arises in such environments because, due to random arrival of tasks and their random CPU service time requirements, there is good possibility that several computers are heavily loaded (hence suffering from performance degradation), while others are idle or lightly loaded (Figure 4.7). Clearly if the work load at some computers is typically heavier that at others, or if some processors execute tasks at slower rate than others, this situation is likely to occur often.

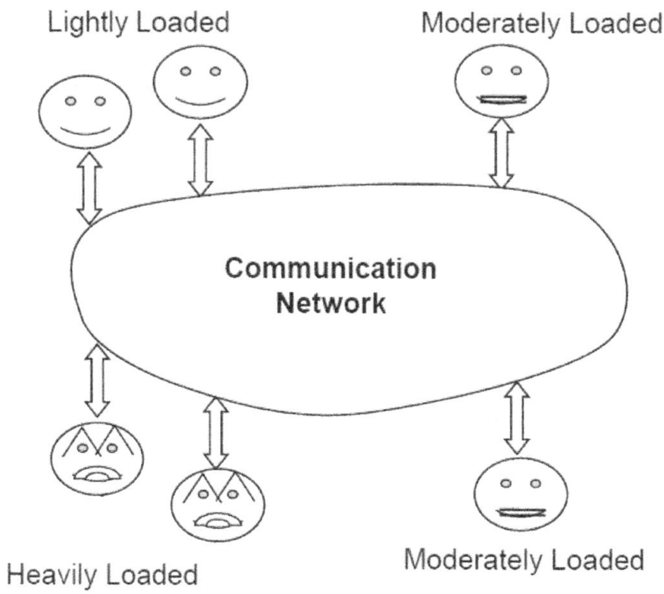

Figure 4.7: Load distribution

Therefore, even in homogeneous distributed system, system performance can be improved by appropriately transferring the load from heavily loaded computers to lightly loaded computers. This raises the following two questions. First what is meant by performance? One widely used performance metric is average response time of tasks. The response time of a task is the length of time interval between its origination to its completion. Minimizing the average response time is the goal of load distribution. Second, what constitutes a proper characterization of load at node? Defining a proper load index is very important as load distributing decisions are based on the load measured at one or more nodes. Also it is crucial that the mechanism used to measure load is efficient and imposes minimum overhead.

Issues in Load Distribution

The several issues in load distribution are discussed following:

Load: It has been seen that resource queue length and particularly the CPU queue length are good indicators of load because they correlate well with the task response time.

Classification of Load Distributing Algorithm: The Basic function of load distributed algorithm is to transfer load from heavily loaded computers to lightly loaded computers. Load Distributing algorithms are characterized as follows:

- Static
- Dynamic
- Adaptive

Dynamic Load distributing algorithms use system state information (the loads at nodes) to make load distributing decisions, while static algorithms make no use of such information. In static algorithms, decisions are hard-wired in the algorithms using priori knowledge of systems. Dynamic load distributing algorithms have the potential to outperform static load distributing algorithms because they are able to exploit short term fluctuations in the system state to improve performance. However, Dynamic load distributing algorithms entail overhead in collection, storage, and analysis of system state information. Adaptive load distributing algorithms are special class of

Dynamic load distributing algorithms in that they adapt their activities by dynamically changing the parameters of algorithm to suit the changing system state. For example, a Dynamic load distributing algorithm may continue to collect the system state irrespective of system load. An adaptive algorithm, on the other hand, may discontinue the collection of system state if the overall system state is high to avoid imposing additional overhead on the system.

Load Balancing vs Load Sharing: Load distributing algorithms can further be classified as *load balancing* and *load sharing* algorithms, based on their load distributing principle. Both types of algorithms try to reduce the likelihood to an *unshared state.* A unshared state is the state where a node lies idle while at the same time tasks contend for service at another node by transferring tasks to lightly loaded nodes.

Load balancing algorithms try to equalize loads at all computers. Because a load balancing algorithm transfers tasks at higher rate, higher overhead incurred by load balancing algorithms may outweigh this potential performance improvement.

Task transfers are not instantaneous of communication delays and delays that occur during the collection of task state. Delays in transferring a task increase the duration of unshared state as an idle node must wait for the arrival of transferred task. To avoid the lengthy unshared states, *anticipatory* task transfers from overloaded node to the nodes that are likely to become idle can be used. Anticipatory transfers increase the task transfer rate of load sharing algorithm. In this sense, load balancing can be considered a special case of load sharing, performing a particular level of anticipatory task transfer.

Preemptive vs Non Preemptive Transfers: Preemptive task transfer involves the transfer of task that is partially executed. This transfer is expensive as the collection of task's state (which can be quite large and complex) can be difficult. Typically a task state consist of a virtual memory image, a process control block, unread I/O buffers and messages, file pointers etc.

On the other hand, Non preemptive task transfers, involve the transfer of tasks that have no begun execution and hence do no require the transfer of the task's state. In both type of transfers, information about the environment in which the task will execute must be transferred to the receiving node.

Components of Load Distributing Algorithms

Typically, a load distributing algorithm has following four components:

1. ***Transfer Policy*:** Transfer policy determines whether a node is in a suitable state to participate in task transfer. A large number of transfer policies that have been proposed are *threshold* policies. Thresholds are expressed in unit of loads. When a new task originates at node, and load at that node exceeds the threshold, the transfer policy decides that that node is a sender. If the load at a node falls below the threshold, the transfer policy decides that the node is receiver for a remote task.
2. ***Selection Policy:*** A selection policy determines which task should be transferred, once the transfer policy decides the node is sender. Whenever the selection policy fails to find a suitable task to transfer, node is no longer considered a sender until transfer policy decides the node is sender again.

   The simplest way is to select newly originated task that have caused the node to become a sender by increasing the load at the node beyond the threshold. Such tasks are relatively cheap to transfer, as the transfer is non preemptive.

   A basic criterion that a task selected for transfer should satisfy is that the overhead incurred in the transfer of the task should be compensated for by the reduction in the response time realized by the task In general, long-lived tasks satisfy this criterion.

   There are other factors to consider in selection of task. First the overhead incurred by the transfer should be minimal. For example, a task of small size carries less overhead. Second, he number of location independent system calls made b he selected task should be minimal. Location dependent system calls must be executed at the node where the task originated because they use resources such as windows, or mouse, that only exist at that node.

3. ***Location Policy:*** The location policy determines to which node the task selected for transfer should be sent. A widely used method for finding a suitable node is through *Polling.* I the polling, a node polls another node to find out whether it is suitable node for load sharing. Node can be polled either serially or in parallel (by multicasting). A node can be selected for polling either randomly, based on the information collected during the previous polls, or on the nearest neighbour basis.
4. ***Information Policy:*** The information policy is responsible for deciding when information about the states of other nodes in the system should be collected, where it should be collected from, and what information should be collected. The most policies are one of he following here types:
   - *Demand Driven:* In this class of policy, a node collects the state of other nodes only when it becomes either a sender or receiver, making it suitable candidate to initiate load sharing. Demand driven policies can be *sender-initiated, receiver-initiated or symmetrically initiated.* In sender initiated policies, sender looks for receivers to transfer their load. In the receiver initiated policies, receiver solicit load from senders. A symmetrically initiated policy is combination of both, where load sharing actions are triggered by the demand for extra processing power or extra work.
   - *Periodic:* In this class of policy, nodes exchange load information periodically. Based on information collected, the transfer policy at a node may decide to transfer job.
   - *State Change Driven:* In this class of policy, nodes disseminate state of information whenever their state changes by a certain degree. A state change driven policy differs from a demand driven policy in that it disseminates information about the state of a node, rather than collecting information about other nodes.

4.3.2 Stability

Stability can be understood in two perspectives as follows:

1. ***The queuing Theoretic Perspective:*** When the long term arrival rate of work to a system is greater than the rate at with the system can perform work, the CPU queue grows without bound. Such a system is termed *unstable.* For example, consider a load distributing algorithm performing excessive message exchanges to collect state information. The sum of the load due to the eternal work arriving and the load due to overhead imposed by the algorithm can become higher than the service capacity of the system, causing system instability. A load distributing algorithm can be said to be effective under a given set of conditions if it improves the performance relative to that of system not using load distributing algorithm.
2. ***The Algorithmic Perspective:*** If the algorithm can perform fruitless actions infinitely with finite probability, the algorithm is said unstable. For example, consider *node thrashing.* The transfer of task to a receiver may increase the receiver's queue length to the point of overload, necessitating the transfer of the task to yet another node. This process may be repeated infinitely. In this case, a task is moved from one node to another in search of a lightly loaded node without ever receiving service.

4.3.3 Task Migration

Task migration facilities allow preemptive transfers. At this time, it is necessary to make a distinction between *task placement and task migration.* Task placement refers to the transfer of the task that is yet to begin execution to a new location and starts its execution there. The task migration refers to the transfer of the task that has already begun execution to a new location and continuing its execution there. To migrate a partially executed task to a new location, the task's state should be made available at new location.

The general steps involved in task migration are:

- **State transfer:** Task's state includes such information as the content of registers, the task's stack(ready or blocked), virtual memory address space, file descriptor, any temporary files ha he ask might have created, and buffered messages. In addition, the current working directory, resource usage statistics, references to the children processes etc. may be maintained by the kernel as part of task state. The task is suspended at some point during the transfer so that the state does not change further, and then the transfer of the task's state is completed.

- **Unfreeze:** The task is installed at new machine and put in the ready queue so that it can continue executing.

4.3.4 Fault Tolerance

The disruptions caused during failures can be especially sever in many cases, such as on line transaction processing, process control and computer based communication user communities. To avoid disruptions due to failures and to improve availability, systems are designed fault tolerant.

Since the fault tolerant systems must behave in specified manner in the event of failure, it is important to study the implications of certain types of failures.

- **Process Death:** When a process dies, it is important that the resources allocated to that process are recouped, otherwise they may be permanently lost. Many distributed systems are structured along the client server model in which a client requests a service by sending a message to a server. If the server process fail, it is necessary that the client machine be informed so that the client process, waiting for a reply can be unblocked to take suitable action. In the same way if client process dies after sending a request to a server, it is imperative that the server be informed that the client process no longer exists. This will facilitate the server in reclaiming any resources it has allocated to the client process.
- **Machine Failure:** In case of machine failure, all the processes running at that machine will die. As far as the behaviour of client process or a server process is concerned, there is not so much deference in their behaviour in the event of machine failure or process

death. The only difference lies in how the failure is detected. In the case of process death, other processes including the kernel are active. Hence, a message stating that the process is died can be sent to inquiring process. On the other hand, an absence of any kind of message indicates either process death or a failure due to machine failure.

- **Network Failure:** A communication link failure can partition a network into subnets, making it impossible to communicate with another machine in a different subnet. A process can not tell really tell the difference between machine and communication link failure, unless the underlying communication network can recognize a machine failure.

A system can be designed to be fault tolerant in two ways.

- Mask failure
- Well defined failure behaviour.

When a system is designed to mask failure, it continues to perform its specified function in the event of failure. A system designed for well defined behaviour may or may not perform the specified function in the event of failure, however, it can facilitate actions suitable for recovery. An example of well defined behaviour during a failure is: the changes made to a database by transaction are made visible to other transactions only if the transaction successfully commits. If the transaction fails, the changes made to database by the failed transaction are not made to visible to the other transactions, thus not affecting those transactions.

One key approach used to tolerate failures is *redundancy.* In this approach a system may employ a multiple number of processes, employ a multiple number of hardware components, multiple copies of data etc. each with independent failure modes (i.e. failure of one component does not affect the operation of other components).

4.3.5 Security

Security and protection deal with the control of unauthorized use and access to the hardware and software resources of computer system. Business organizations and government agencies heavily use computers to store information to which unauthorized access must be prevented. For example, in business organizations, this information includes financial or personnel records, monetary transactions, legal contracts, payrolls, product information future planning and strategies etc. Clearly, an unauthorized use of company's confidential information can have catastrophic financial consequences and unauthorized use of secret information of the government can have serious implications for the security of that nation. Therefore, with the widespread use of computers in business and government organizations, the security and protection of computer systems have become extremely important factors. Computer system security can be divided into two parts:

- ***External Security:*** External security as also called physical security. It deals with regulating access to the premises of the computer system, which includes the physical machine(hardware, disks, tapes, power supply, air conditioning), terminals, computer console etc. External security can be enforced by placing a guard at door, by giving a key or secret code to authorized person etc.
- ***Internal Security:*** Internal security deals with the access and use of computer hardware and software information stored in computer system. Aside from external and internal securities, there is an issue of authentication by which a user "logs into" the computer system to access the hardware and the software resources.

Access Matrix Model: Access matrix model was first proposed by Lampson. The model consists of following three components.

- ***Current Objects:*** Current objects are the finite set of entities to which access to be controlled. The set is denoted by 'O'. A typical example of an object is a file.
- ***Current Subjects:*** Current subjects are the finite set of entities that access current objects. The set is denoted by 'S'. A typical example of a subject is process.
- ***Generic Rights:*** A finite set of generic rights, $R=\{r_1, r_2, r_3......, r_m\}$, gives various access rights that subject can

have to objects, Typical example of such rights are read, write, execute, delete etc.

Protection state of the system: The protection state of the system is represented by a triplet (S,O,P), where S is the set of current subjects, O is the set of current objects, and P is the matrix called access matrix with the row of every current subject and a column for every current object. A systematic diagram of an access matrix is shown figure 4.8:

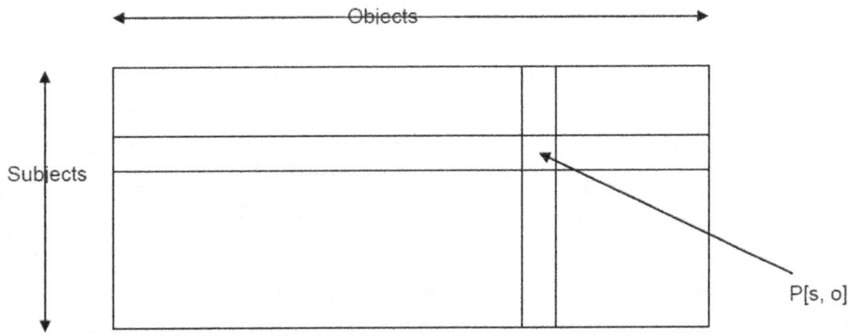

Figure 4.8: Protection state

Note that the access matrix P itself is a protected object. Let the variables s and o denote a subject and an object respectively, Entry P[s, o] is a subset of R and denotes the access rights which subject s has to object o.

Enforcing a Security Policy

A security policy is enforced by validating every user access for appropriate access rights. Every object has a monitor that validates all access to that object in the following manner:

- A subject s requests an access α to the object o.
- The protection system presents triplet (s, α, o) to the monitor of o.
- The monitor looks in to the access rights of o. If α ∈ P[s, o], then the access is permitted else it is denied.

EXERCISE

Q1. Discuss the architecture of Distributed File System

Q2. What is the work of cache in file system?

Q3. Write Short notes on following:

a. File Server

b. Name Server

c. Encryption

Q4. What procedure is used for mounting a name space in Distributed File System?

Q5. What do you understand by cache consistency problem? Discuss the approaches for solving this problem.

Q6. Define the process of file replication for the file availability.

Q7. Discuss the significance of Shared memory in distributed environment.

Q8. Differentiate between message passing and distributed shared memory.

Q9. Discuss the architecture of Distributed shared memory. Aldo discuss the Migration Algorithm, Read Replication Algorithm and Full Replication Algorithm for implementing DSM.

Q10. How the distributed shared memory can be implemented. Give any one algorithm.

Q11. What do you mean by memory coherence? Discuss the different types of consistency.

Q12. Discuss different writing policies concerned with updating the blocks from client to server.

Q13. What do you mean by location transparency?

5

Communication Protocols

5.1 INTRODUCTION

The communication protocols are used for reliable exchange of data between two sites. So the main functions of these protocols is **reliable data transmission**. The reliable data transmission includes the:

- Repeated sending of messages that are lost
- Rejecting or correcting messages that are garbled
- Discarding duplicate messages

To do so the protocols maintain state information, recording which data has already been sent, which data has been certified to be received and so on.

For two entities to successfully communicate, they must "speak the same language". What is communicated, when it is communicated, and how it is communicated must confirm to some mutually acceptable set of conventions between the entities involved. The set of conventions is referred as communication protocol, which may be defined as a set of rules governing the exchange of data between two entities.

The key elements of the protocols are:

- **Syntax:** Includes such things as data format, coding and signal format
- **Semantics:** Includes control information for co-ordination and error handling.

- **Timing:** Includes speed, matching and sequencing.

5.2 BALANCED SLIDING WINDOW PROTOCOL

This protocol is used for exchange of information between stations that are directly connected through a line.

Let us see how this protocol works. Assume A and B are the sites connected via full duplex link. If site B allocates buffer space for n frames, means A is allowed to send n frames without waiting for any acknowledgement. To keep track of which frames have been acknowledged and how many frames have been sent, each is labeled with a sequence number.

B acknowledges a frame by sending on acknowledgement that includes the sequence number of next frame expected. This acknowledgement also announces that B is prepared to receive next n frames, beginning with number specified.

As the sequence number occupies a field in frame, it is clearly of bounded size. Assume if 3 bit sequence number is used, then frames may be numbered from 0 to 7.

The Sliding window protocol is described below:

At the Sender's end:

- A sliding window may have maximum 2^n frames (n is bits sequence number).
- The frames in window indicate that they are to be sent.
- As the frame is sent, that frame is excluded from window, thus window shrinks.
- As the acknowledgement is received, the window grows upto maximum limit from the sequence number specified in acknowledgement.

At the Receiver's end:

- The frames in the window at receiver end denote that receiver is ready to receive those frames.
- As soon as a frame is received, that frame is excluded from window thus window shrinks.
- When the receiver sends the acknowledgement, the windows grow upto the maximum possible size from the sequence number specified in acknowledgement.

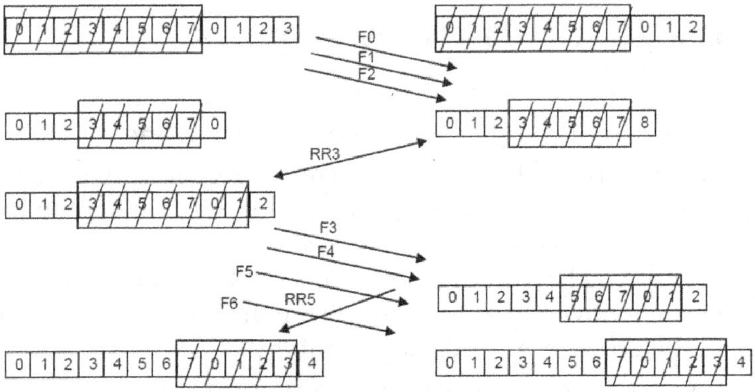

Figure 5.1: Sliding window protocol

Note: An acknowledgement message RR3 (ready to receive 3) indicate "I have received all frames up through frame number 2 and am ready to receive frame number 3; in fact I am ready to receive seven frames beginning with frame number 3."

5.3 ROUTING ALGORITHMS

A site in general is not generally connected directly to every site in distributed system. A site can send packets of information directly to a subset of sites called neighbors.

Routing is the term used to describe the decision procedures by which a site selects one or more of its neighbors to forward a packet on its way to on ultimate destination. The objective in designing a routing algorithm is to guarantee a decision procedure to this function and guarantee delivery of each packet.

For the routing decisions each site must maintain the information about the topology of networks. Such information is maintained in routing tables. So the routing algorithms must ensures the following:

- **Table Computation:** The routing tables must be computed when network is initialized and must be brought up to date if topology of network changes.

- **Packet Forwarding:** When a packet is to be sent through network, it must be forwarded using routing tables.

Objectives of Good Routing Algorithms

1. Correctness : The algorithm must deliver every packet offered to network, to its ultimate destination.

2. Efficiency : The algorithm must send packets through "good" paths. The path that suffers only a small delay and ensures high throughput, is considered as good.

Note: The routing algorithm is said optimal if it uses the best path.

3. Complexity : The complexity of the routing algorithm should be as less as possible. The aspect of complexity includes:

- How fast the routing decisions can be made.
- How fast a packet can be made ready for transmission.
- How much storage space the routing table takes.

4. Robustness: Robust means strong and not likely to fail in a situation. The robustness of routing algorithm means in the case of topological change, the algorithm updates the routing tables in order to perform the routing function in modified network.

Adaptiveness: The algorithm balances the load of channels and nodes by adopting the tables in order to avoid paths through channels or nodes with high load and preferring the channels or nodes with currently light load.

Fairness: The algorithm must provide service to every user in same degree.

Criteria for Optimal Routing Algorithm

As we know that network is represented as a graph, where the nodes of the graph are the nodes of the network, and there is an edge between two nodes if they are neighbours (i.e. they have a communication channel between them). The optimality of an algorithm depends on what is "best" path in the graph; there are several notions of what is "best", each with its own class of routing algorithms:

1. **Minimum Hop:** The cost of using a path is measured as the number of hops (traversed channels or steps fro node to node) of the path. A minimum hop routing algorithm uses a path with the smallest possible number of hops.
2. **Smallest Path:** Each channel is statically assigned a (non-negative) weight and the cost of a path is measured as the sum of the weights of the channels in the path. A shortest part algorithm uses a path with lowest possible cost.
3. **Minimum Delay:** Each channel is dynamically assigned a weight depending on the traffic on the channel A minimum delay algorithm repeatedly revises the table in such a way with a (near) minimal total delay are always chosen.

Destination Based Routing

The routing decisions made when forwarding a packet is usually based only on destination of packet and is independent of the original sender of packet.

A path from a source to destination is called optimal if there exists no path between them with lower cost.

Note: The optimal path is not always unique, there may exist different paths with same minimal cost.

Pre Requisites for Destination Based Routing

The cost of sending a packet via a path is actual utilization of path. The cost of using a path P is being denoted here as a function of path C(R).

The cost of concatenation of two paths equals to sum of the costs of concatenated paths.

$C(U_1, U_2, U_3 \ldots\ldots U_k) = C(U_1, U_2, \ldots\ldots U_i) + C(U_i, \ldots\ldots U_k)n$

Note: The cost of forwarding a packet along a Path P is computed by considering the distance, bandwidth, average traffic, communication cost, measured delay etc.

The destination based routing uses the optimality principal-which is as follows:

Optimality Principal: If a node J is an optimal path from node I to K, then optimal path from J to K also falls along same route.

Proof: Let r_1 be the part of that optimal path and connects the I to J.

Let the rest of that optimal path is r_2, which connects J to K.

If a path better than r_2 exists from J to K, it must be concatenated with r, to improve the optimal path from I to J. And that is contradicting statement so r_1 r_2 is optimal.

Thus by using the optimality principal, in the destination based routing, the optimal path from all sources to the destination are searched, which results in a tree rooted at destination. Such a tree is called optimal sink tree.

Consider the following network:

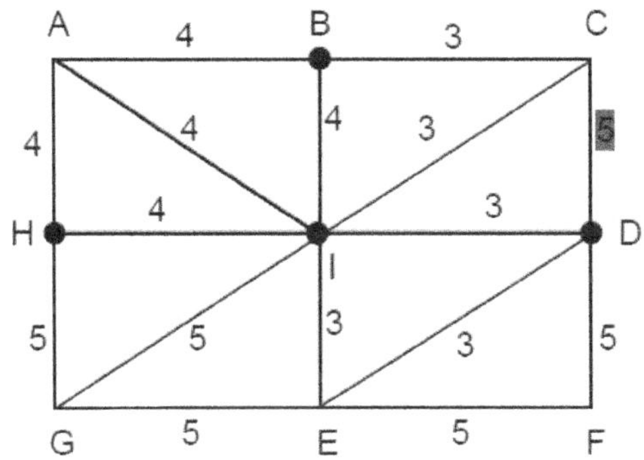

If the destination is B the optimal sink tree will be as:

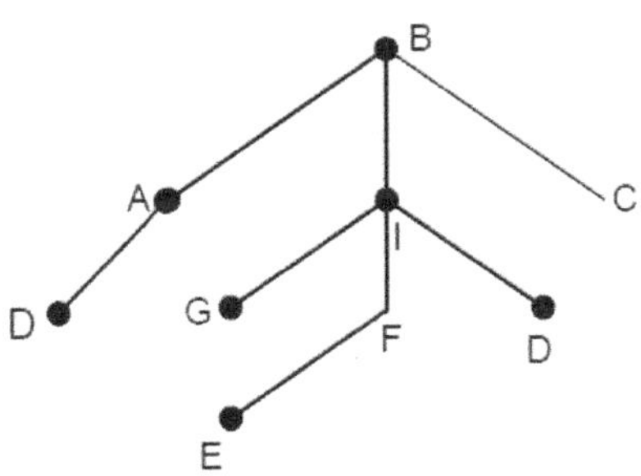

The existence of optimal sink tree implies that it is no compromise to optimality, if following packet forwarding mechanism is used.

(when packet with destination D was received)
if d = 4
then deliver the packet locally
else send packet to table_lookup$_u$ (d)
Here table_lookup$_u$ (d) returns the father of u in sink tree.

Deadlock-Free Packet Switching

First we should know what is packet switching.

In the packet switching data is transmitted in short packets. A typical upper bound on packet length is 1000 bytes. If a source has longer message to send, the message is broken up into a series of packets. Each packet contains a portion of user's data plus some control information. The control information, at a minimum, includes the information that the network requires in order to be able to route the packet through the network and deliver it to intended destination. This packed is stored at each node before being forwarded to next node on the path to their destination.

Each node of the network reserves some buffer space for this store-and-forward purpose. As the amount of buffer space is finite in each node, a situation may occur where no packet can be forwarded because all buffers in next node are occupied.

The situation becomes the worst case if a group of packets can never reach to their destination because they all are waiting for the use of buffer currently occupied by another packet in the group. Such a situation is known as store-and-forward deadlock.

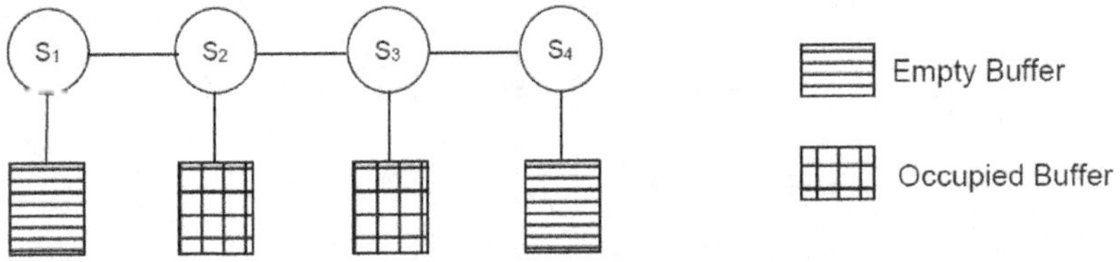

In above example suppose buffer's capacity is exactly the size of one packet. If S_1 has sent a packet to S_2 with destination S_4, and S_4 sends a packet to S_3 with destination S_1 none of the packets stored in buffer of S_2 and S_3 can be forwarded to its destination.

So in the design of packet switching networks, following two approaches are introduced which avoids the possibility of store and forward deadlock, by imposing restrictions on when a packet can be generated or forwarded.

- Structured Approach
- Unstructured Approach

Structured Solution

In this approach Buffer Graph is constructed. A ***Buffer Graph*** BG(B , E) is a directed graph on the buffers of network such that:

- BG is acyclic
- B={b_1, b_2 ----b_s} where b_1, b_2 .....b_s are the buffers in nodes of network
- If b_1, b_2 ----b_s is a path in BG, then. If u_i is the node where buffer b_i resides, then u_0, u_1, . . .u_s is a sequence of node such that for each i< s either $u_i u_{i+1}$ Є E or $u_i = u_{i+1}$.
- bc Є BE implies that b and c are buffers in the same node or buffers in the two node connected by a channel.
- Let p be a packet in the node u with destination v. A buffer b is *suitable* for p if there is a path in BG from b to a buffer c in v. Such a path is designated as ***Guaranteed Path*** . nb(p,b) denotes the next buffer on the guaranteed path and fb(p) denotes the first buffer for packet p.

Structured Solution is implemented as follows:

The Generation is packet p in u is allowed if and only if the buffer fb(p) is free. If the packet is generated it is placed in this buffer.

The forwarding a packet p from a buffer in u to a buffer in w is allowed if and only if nb(p,b) is free. If the forwarding takes place p is placed in nb(p,b).

Unstructured Solution

Unlike to structured solution this solution do not prescribe in which buffer a packet must be placed and uses simple local information such as hop count or the number of occupied buffers in a node. This solution is implemented by ***Forward-Count*** and ***Backward-Count*** Controllers.

Forward-Count Controller:

Forward-Count Controller accepts a packet if the node contains more free buffers than the packet has hops to go i.e the controller accepts a packet p in node u if and only if $s_p < f_u$

Where

s_p : Number of hops the packet has to make to its destination

f_u : Number of free buffers in u.

Forward-Count Controller:

This controller accepts a packet on the basis of number of steps a packet has already made. This Controller accepts a packet p in node u if and only if $t_p > k - f_u$

Where

t_p : Number of hops the packet has made from its source

k : Total number of hops

f_u : Number of free buffers in u.

5.4 WAVE AND TRAVERSAL ALGORITHM

Computer networks are the natural application field of distributed algorithms. Distributed algorithms are designed for an interconnected collection of autonomous processes (or processors), cooperating to solve given problems.

In the design of distributed algorithms, very general problems appear frequently as sub tastes. These elementary subtasks include.

- Broadcasting of Information
- Achieving Global Synchronization
- Executing Distributed Computation (each process holds a part of input)
- Mutual Exclusion, etc.

Wave algorithms (sometimes referred as total algorithms) are widely used as the basis to solve many network control problems. “A wave algorithm is a distributed algorithm where the participation of all the processes in network is required before a particular event, called a decision, is taken.”

Definition of Wave Algorithm

Consider a computation C

(Each computation is collection of events partially ordered by casual precedence relation μ)

- Set of all process is denoted by P.
- The number of events in computation denoted as |C|.
- The subset of events of computation C that occur in process p is denoted as C_p.

A wave algorithm is a distributed algorithm that satisfies the following three requirements:

Termination: Each Computation is finite

$\forall C: |C| \mu \infty$

Decision: Each computation contains at least one decide event.

$\forall C: \exists e \in C$: e is decide event.

Dependence: In each computation, each decide event is casually preceded by an event in each process.

$\forall C: \forall e \in C$: (e is decide event $\Rightarrow \forall q \in P \exists f \in C_q: f \mu e$)

A wave algorithm exchanges a finite number of messages and then makes a decision, which depends casually on some event in each process.

In the computation of algorithm some processes are known as initiators (also known as starters) and some are known as non-initiators (also known as followers).

- A process is initiator if it starts the execution of its local algorithm spontaneously. The first event of initiator is send event.
- A non-initiator is involved in algorithm only when a message of algorithm arrives and triggers the execution of process algorithm. The first event of non-initiator is receive event.

Variety of Wave Algorithms

The variety of wave algorithms exists because algorithms may differ in many respects. The list of aspects in which wave algorithms differ form each other is given below:

Centralization: An algorithm is called centralized if there must be exactly one initiator in each computation. These algorithms are also called single-source algorithms.

An algorithm is called decentralized if the algorithm can be started spontaneously by an arbitrary subset of processes. These algorithms are also known as multiple source algorithms.

Topology: An algorithm may be designed for a specific topology such as ring, tree etc.

Classification of Wave Algorithms

Wave algorithms include token circulation (TC) and propagation of information with feed back (PIF).

- The purpose of TC scheme is to implement a token circulation scheme where the token, initiated by a process (referred as root) is passed through every process in network, is returned to root.
- The PIF scheme can be informally described as follows: "Root initiates a wave by broadcasting a message m. The wave terminates when all processes have acknowledged to root, the receipt of m.

Note: The TC and PIF schemes are useful when broadcasting of messages is required. Wave and Traversal algorithms are also useful for synchronization. Consider the following scenario.

In each process q an event a_q must be executed, in some processes an event b_p must be executed such that the execution of all a_q event must have taken place before any of the b_p events is executed.

For such synchronization problem, event b_p can be considered as decide event in wave algorithm.

General Wave Algorithms

- Ring algorithm
- Tree algorithm
- Echo algorithm
- Polling algorithm
- Phase algorithm
- Finn's algorithm

Ring Algorithm: This is the centralized approach and uses the TC scheme. This is wave algorithm for ring network.

The initiator sends a message <tok:> (called as token) along the cycle, each process passes it on and when it returns to initiator, the initiator decides

For the initiator

```
begin
    send < tok > to next_p
    receive < tok >
    decide
end.
```

For non-initiators

```
begin
    receive < tok >
    send < tok > to next_p
end.
```

2. The Tree Algorithm: This wave algorithm is for tree network or it can also be used in arbitrary network if spanning tree is available. Thus is de-centralized approach.

It is assumed that all leaves of tree initiate the algorithm. Each process sends exactly one message. If a process has received a message via each of its incident channels except one, the process sends a message via remaining channel. If a process has received a message via all of its channels it decides.

Boolean rec_p [q] for each $q \in Neigh_P$

(initialized with false means P has not received message from q)

```
begin
    while # {q : rec_p [ q ] is false} > 1
        begin
            receive < tok > from q ; rec_p [ q ] = true
        end
(Now there is one q_0 with rec_p [ q_0 ] = false)
```

```
            send < tok > to q0
            receive < tok > from q0
            rec_p [ q0 ] = true
            decide
(Inform other processes of decision)
    forall q ∈ Neigh_p
            send < tok > to q
end
```

Example:

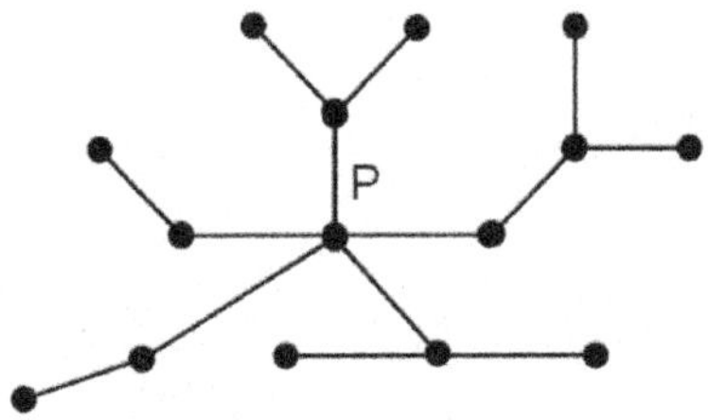

Given Spanning Tree

The decomposition of spanning tree if process P decides first.

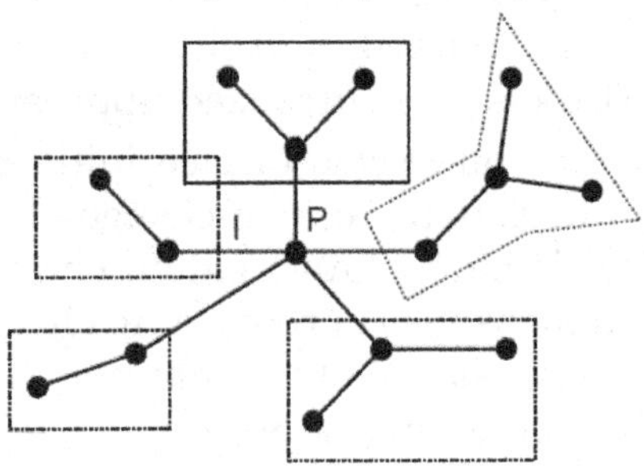

3. Echo Algorithm: The echo algorithm is centralized wave algorithm for network of arbitrary topology. It uses the propagation of information with feedback (PIF) scheme.

In this algorithm, initiator sends message to all its neighbors. Upon receipt of message, a non-initiator forwards message to all its neighbors except the one from which the message was received and makes the process from which it has received the message, its father. When a process has no neighbor except its father, it sends echo message immediately. When non-initiator has received echo message from all its neighbors it sends echo message to its father. When initiator has received echo message from all its neighbors it decides.

```
int rec_p      (counts number of received messages)
Process father_p
```

For initiator

```
begin
    forall q ∈ Neigh_p
            send < tok > to q
    while rec_p < # Neigh_p
            begin
                    receive < tok > ; rec_p = rec_p +1
            end
    decide
end
```

For non-initiator

```
begin
    receive < tok > from neighbor q;
    father_p = q
    rec_p = rec_p +1
    forall q ∈ Neigh_p : q.# father_p
            send < tok > to q
    while rec_p < # Neigh_p
            begin
                    receive < tok >
                    rec_p = rec_p + 1
            end
    send < tok > to father P
end
```

4. Polling Algorithm: This algorithm is used in star network in which initiator is in center.

The initiator asks each neighbor to reply with message and decides after receipt of all messages.

```
int rec_p
For initiator
begin
    forall q ∈ Neigh_p
            send < tok > to q
            while rec_p < # Neigh_p
            begin
                    receive < tok >
                    rec_p = rec_p + 1
            end
            decide;
end

For non-initiators
begin
    receive < tok > from q
    send < tok > to q
end.
```

5. Phase Algorithm: It is a decentralized algorithm for network of arbitrary topology. This can be used as wave algorithm for directed networks. The algorithm requires that the process know the diameter D of the network. The algorithm is also correct if the process uses instead of D, a constant D' larger than the network diameter. Thus in order to apply this algorithm it is not necessary that diameter is known exactly; it suffices if the upper bound (i.e. N-1) on the network diameter is known.

As it can be used in arbitrary directed networks where channels can carry messages in one direction only, the neighbours of node p are:

In-neighbour : Process that can send messages to p

Out-neighbours: Process to which p can send messages.

In the phase algorithm each process sends exactly D messages to each out-neighbour. Only after i messages have been received from each in-neighbour, (i + 1)th message is sent to each out neighbour.

6. Finn's Algorithm: This is another wave algorithm that can be used in arbitrary directed networks. It does nor require the diameter of the network to be known in advance, but relies on the availability of the unique identities for the processes. Set of processes identities are exchanged in messages, which causes the bit complexity of the algorithm to be rather high.

Process p maintains two set of process identities:

Inc_p : is the set of processes q such that an event in q precedes the most recent event in p

$NInc_p$: is the set of processes q such that for all neighbours r of q ,an event in r precedes the most recent event in p.

This relation is maintained as follows :

Initially $Inc_p = \{p\}$ and $NInc_p = \acute{\emptyset}$. Process p sends message, including Inc_p and $NInc_p$, each time one of he se has increased. When p receives a message, including an *Inc* set and *NInc* set, these received identities are inserted into p's version of these sets. When the p has received from all in-neighbours p is inserted into $NInc_p$. When the two sets become equal p decides.

5.5 ELECTION ALGORITHMS

An algorithm for choosing a unique process to play a particular role is called election algorithm. For example, in a variant of client-server algorithms for mutual exclusion or deadlock detection one server or coordinator is chosen by all processes. An election algorithm is needed to choose which of the process will play the role of coordinator. It is essential that all the processes agree on same choice. Afterwards, if a process that plays the role of coordinator wishes to retire, then another election is required to choose a replacement.

In nutshell Election process can be understood as follows:

- A leader is often needed in distributed systems
  - As controller or coordinator
- We need to elect a leader on startup and when current leader fails
- Assumption
  - Every process knows ID of all the other processes
- Conditions

- Operational process with largest ID wins
- All operational process should be informed of a new leader
- Recovering process can find current leader

Following two algorithm may be used for election process:

Bully Algorithm: This algorithm was devised by Garcia-Molina. In this algorithm election is initiated by any process (P) which notices that coordinator is no longer responding. Concurrent multiple elections are possible.

Algorithm

- P sends ELECTION messages to all process with higher ID
- If no one responds, P wins and becomes coordinator
  - Sends out COORDINATOR messages to all other processes
- If one of higher-ups answers, it takes over. P is done.
- The performance of the Bully Algorithm can be analyzed as:

Best Case

Process with highest ID (among operational ones) starts election

- ELECTION message = 1
- COORDINATOR announcements = N-1

Worst case

Process with lowest ID starts election

- N-1 ELECTION messages + N-1 Acknowledgements = 2(N-1)
- Machine 2 starts ELECTION = 2(N-2)
- Machine N – 1 = 2
- COORDINATOR announcements = N-1
- Total: $O(N^2)$

Ring Algorithm

Assumption

- Processes are physically or logically ordered and construct a ring
- Each process knows its successor (and all others)

Algorithm

- Initiator Send ELECTION message with its ID to its successor
    - If successor is down, it skips over to the next node in the ring
- When ELECTION message is received
    - If it contains the receiver's ID, decides who wins election and convert it to COORDINATOR message and send them to its successor
    - If not, put its own ID and send over to it successor

Variations of Ring Algorithm

- Putting all IDs in ELECTION message is not desirable
    - Keep only max ID among the processes which have been seen so far.
- How can it be known that all live processes have been checked?
    - Receiver's ID is found in the ELECTION message
- How to absorb unnecessary messages from concurrent elections?
    - Identify another election
- Sees RCV_ID < ID while it is currently participating another election. (why?)
    - Then, when concurrent election is redundant, remove it.

Another Ring Algorithm

1. Initially, every process is non-participant
2. Election initiator
    - marks itself participant
    - Sends ELECTION message with its ID to next process

3. Process P receives ELECTION
    - If P_ID < RCV_ID (RCV_ID = ID in ELECTION message)
        - P forward message to next process
        - P marks itself as participant (possibly, already marked)
    - If P_ID > RCV_ID
        - If P is not participant, P replaces ID in ELECTION with P_ID
        - If P is a participant, ignore ELECTION message
    - IF P_ID = RCV_ID
        - P becomes coordinator (why?)
        - Mark itself as non-participant
        - Send out COORDINATOR announcement with its ID to next process
4. Process Q receives COORDINATOR message (Q≠P)
    - P marks itself as non-participant
    - Sets coordinator = RCV_ID (ID in COORDINATOR message)
    - Forwards the message to next process
    - IF Q = P, ignore the message

EXERCISE

Q1. What are the objectives of a Routing Algorithm? Discuss the criteria for selecting optimal routing algorithm. Discuss the Destination Based routing.

Q2. Discuss the followings in short:

a. Wave Algorithms

b. Deadlock free packet switching

Q3. How the Election algorithms are useful in Distributed Systems. Discuss the Bully Algorithm for election of successor.

Made in the USA
Monee, IL
13 September 2026

62981862R00095